Duncan Poore has been Professor of Botany in Malaya,
Scientific Director and, later, Director General of the IUCN,
Professor of Forestry and Director of the Commonwealth
Forestry Institute of Oxford. He was a member of the
Nature Conservancy Council and Chair of its Advisory
Committee on Science. He has also been a specialist adviser,
on aspects of forestry in Britain, to the House of Lords Select
Committee on Science and Technology. He is a consultant
in land use questions and Senior Consultant to IIED.

Simon Rietbergen is a Research Associate in the Forestry
Programme of IIED.

Dr Timothy Synnott, Peter Burgess and John Palmer are
consultants in tropical forestry.

No Timber Without Trees

Sustainability in the Tropical Forest

by Duncan Poore
Peter Burgess
John Palmer
Simon Rietbergen and
Timothy Synnott

A Study for ITTO

Earthscan Publications Ltd   London

This book is based upon a study made for the International Tropical Timber Organization. Without ITTO, it would not have been possible to write it. The International Institute for Environment and Development (IIED) wishes to record its deep appreciation to the Executive Director. The views expressed are those of the authors and do not necessarily represent the attitudes or policy of ITTO.

First published 1989 by
Earthscan Publications Ltd
3 Endsleigh Street, London WC1H 0DD

Copyright © 1989 by International Tropical Timber Organization

British Library Cataloguing in Publication Data
No timber without trees.
  1. Tropical rain forests. Management
  I. Poore, Duncan
  634.9'28

ISBN 1–85383–050–X

Production by David Williams Associates, 01-521 4130
Typeset by Rapid Communications Ltd., London WC1
Printed and bound in Great Britain by
Guernsey Press Ltd, Guernsey, C.I.

Earthscan Publications Ltd is a wholly owned and editorially
independent subsidiary of the International Institute for
Environment and Development (IIED)

Contents

Foreword

The sustainable utilization and conservation of the tropical forest is now high on the world political agenda. Central to this is the question whether it is possible (technically, economically and socially) to harvest timber from suitable parts of the forest in a sustainable and environmentally acceptable manner. For it is harvesting of timber that gives governments the main economic justification for keeping forest as forest. If it is possible to do so, the tropical forest may survive; if it is not, the forest is likely to disappear and with it all of the many benefits that the forest provides – not least the supply of beautiful woods that form the raw material of the timber trade.

It was because of the crucial importance of natural forest management that ITTO commissioned the study that forms the core of this book. It reveals a serious situation, one that should stimulate urgent action; but, even more important, a situation in which there is room for considerable hope and a great opportunity.

The authors have, quite deliberately, imposed strict criteria and, according to these, very little management of tropical forest for timber production passes the test. But they have also analysed what is needed to transform the scene – this is the great significance of this study – and it is evident that it *can* readily be transformed, given the right actions. Equally, it could deteriorate rapidly if governments lose faith in forestry as a solution.

The book goes on to place management for timber in the wider context of tropical forest conservation and the environment, and to describe some of the recent international initiatives. I welcome *No Timber Without Trees* because it concentrates on what is being done, and what can be done. We need more of the right action and we need it now. This analysis will certainly help in arguing the political case in a convincing manner.

Dato Dr B.C.Y. Freezailah
Executive Director, International Tropical Timber Organization
Yokohama, June 1989

Preface

The stimulus for this book comes from a study carried out for the International Tropical Timber Organization (ITTO) by the International Institute for Environment and Development (IIED), and the book is largely based on the report which resulted.

ITTO is a relative newcomer to the tropical forest scene and one which has the potential to be very influential in promoting the wiser use of forest lands in the tropics. It is the main outcome of the International Tropical Timber Agreement which came into force on 1 April 1985. This agreement establishes a framework of international co-operation between the nations that produce tropical timber and those that import it, in order to find solutions to the problems facing the tropical timber economy. Among its objectives are: promoting the expansion and diversification of the international trade in tropical timber; improving the structural conditions of the tropical timber market; promoting and supporting research and development to improve forest management and wood utilization; improving market intelligence; encouraging increased and further processing in producer countries; encouraging members to support and develop reforestation and forest management; and improving the marketing and distribution of tropical timber exports. Clauses such as these are not unusual in commodity agreements. What makes this one unique, however, is its concern for the management of the resource. Particularly significant is the final objective (Article 1 (h)): "To encourage the development of national policies aimed at sustainable utilization and conservation of tropical forests and their genetic resources, and at maintaining the ecological balance in the regions concerned."[1]

The purpose of the study carried out by IIED was to examine the management of natural forest[2] for the sustainable production of timber within the producer countries of ITTO: how much was being successfully managed in a sustainable manner (so that its character as forest was preserved and its potential to produce was maintained); where management was succeeding, what local conditions made it successful; and where it had failed, the reasons for its failure. This analysis led directly to recommendations for action. These are addressed to everyone who will listen and who can influence the situation – governments, aid organizations, non-governmental organizations and the international community in general; but they are specially directed to ITTO itself, and it is hoped that they will assist the Committee on Reforestation and Forest Management of ITTO and the International Tropical Timber Council in formulating future policies and priorities.

The subject of the sustainable management of natural forest is of vital concern to both producers and consumers, because most tropical timber still comes from natural forest and it is generally assumed that it will continue to do so. If this assumption is to prove correct, a substantial area of the world's tropical forests will have to be managed for sustainable production. If, on the other hand, it is found that management on the necessary scale is impracticable, the continued flow of tropical timber for domestic use and international trade will depend upon the development of alternative sources of supply.

There are many who believe that this would be disastrous for the tropical forest. Much of the forest occurs in poor countries which need to make the best economic use of all their natural resources. If it is not used for a purpose which is seen to bring economic and social benefit to the country, it will eventually not survive. Nothing is more certain; it will be converted to agriculture or pasture lands, and all too often will end up as waste.

Some of course should be preserved as it is – for the conservation of soil, water supplies and the wealth of living organisms. But, for the remainder, the most obvious economic choice is the production of timber, supplemented where possible by other more specialized forest products, such as rattan, latex, gums, fruits, drugs and so on. It is therefore vital to find means by which those practices of management which are truly sustainable can be adopted throughout all areas of tropical forest, and to do this very soon.

The best available figures show that tropical forest is being destroyed or degraded very rapidly, although there is uncertainty about the exact rates. At the same time, experience in the arid and semi-arid zones, and in many mountain regions of the world, shows that degradation which starts in isolated spots may suddenly accelerate and take off at an unpredictable speed. This is already evident in some parts of the tropical forest: the vast and unprecedented fires in Kalimantan and Sabah are perhaps the most striking manifestation – but not the only one.

The main message of this book, therefore, is urgency. There is no time to lose.

The study draws upon two main sources of information: country visits and consultations with individuals who have a detailed knowledge of the subject. The bulk of the field work was carried out by three consultants, each concentrating on one continent. They visited the producer countries, held discussions and made field visits in them, and organized a round-table discussion in each continent: in Africa (Simon Rietbergen), covering Cameroon, the Congo, Côte d'Ivoire, Gabon, Ghana and Liberia; in Asia (Peter Burgess) – Indonesia, Malaysia (Peninsular, Sabah and Sarawak), Papua-New Guinea, the Philippines and Thailand; and in South America and the Caribbean (Dr Timothy Synnott) – Bolivia, Brazil, Ecuador, Honduras, Peru and Trinidad and Tobago. India, although a producer member of ITTO, unfortunately did not participate in the study; but we understand that all forest reserves in India have recently been withdrawn from timber production.

In addition information has been included on the management of the rain forests in Queensland, Australia (by Duncan Poore), because the experience of this member of ITTO is highly relevant, even though none of the timber produced there enters international trade.

The reports on these country visits were supplemented by discussions with experts in the field, and by an independent commentary on the subject commissioned from Dr John Palmer based on his extensive field experience and knowledge of the literature. Both the Forest Resources Division of FAO and the Co-ordinating Unit of the Tropical Forest Action Plan (TFAP) were consulted during the study.

Although ITTO does not yet include all the countries which possess tropical forest or have experience in its management, the members do contain at least 70 per cent of tropical forest resources; moreover, if there is known to be important experience in non-member countries, this has been taken into account. It is thought, therefore, that this report gives a fair picture of the general situation throughout the tropics.

There have been a number of recent reviews of different aspects of this problem, many of them concentrating on technical and silvicultural aspects. Notable are those by FAO;[3] some parts of the report of the Harvard Institute for International Development (HIID) to ITTO;[4] the report of a workshop of the International Union of Biological Sciences;[5] papers from a symposium at Yale;[6]

an important study on the effect of national policies on forest resources;[7] and three significant studies by IIED.[8] A bibliography of literature on tropical forest management was also produced as part of the report to ITTO.[9] We have not attempted to duplicate all this valuable material but have instead examined the subject from a rather different perspective.

There are ten chapters in the book. In the first we examine the issues involved in natural-forest management for the sustainable production of timber, and the relationship of this to the sustainable provision of other goods and services – a subject which is dealt with in much greater detail in the HIID study (see Note 4). We go on to discuss the place of forest management in the context of policies for the sustainable use and conservation of forest lands in general, and look at the kind of ideal pattern of land use which national policies might take as a model. In the second chapter there is a description of the management of the tropical forests of Queensland, an example of a state in which considerable progress has been made towards applying many of the measures which we have recognized in other countries as essential for successful sustainable management. The third, fourth and fifth chapters deal, respectively, with the present situation in Africa, in South America and the Caribbean, and in Asia. These chapters vary in their approach and tend to concentrate on different aspects, because the situations in the three continents differ greatly. Chapter 6 is John Palmer's assessment. The next chapter, the seventh, is the kernel of the report; it analyses the regional and country reports and draws conclusions about the present state of sustainable forest management.

Chapter 8 is devoted to recommendations for action, carefully selected to have the greatest possible impact on what are seen to be the main obstacles to progress. The last two chapters review recent developments and take a speculative look at the future.

I quote now from the Executive Summary of the report to ITTO:

The main conclusions are sufficiently alarming to warrant urgent action. They are:

(1) The extent of tropical moist forest which is being deliberately managed at an operational scale for the sustainable production of timber is, on a world scale, negligible.

(2) Many countries have the intention, expressed in their

legislation, to manage sustainably; and in a number of them partial forms of sustainable management are being practised either deliberately [or by chance].

(3) There seems at last to be some awareness in most of the producer countries, particularly among their foresters, that action is needed. Many countries, especially with the assistance of FAO and through the Tropical Forestry Action Plan, are now taking some of the necessary steps to attain sustainable management.

(4) Nevertheless, progress in establishing stable sustainable systems is still so slow that it is having very little impact on the general decline in quantity and quality of the forest.

(5) Comprehensive and urgent measures are absolutely necessary if the tropical timber trade is to continue in the long term to handle material which even approaches the quantity and quality that it has become accustomed to.

(6) The future existence of large areas of tropical forest, perhaps even the majority, and of the highly significant ancillary goods and services of the forest, depends equally on the establishment of sustainable systems of management, many of which must have timber production as their basis.

This description of the present situation is followed by a critical analysis of the reasons for the widespread failure of the many valiant attempts to manage natural forests for the sustained production of timber. This leads to a list of the necessary conditions for successful management.

NECESSARY CONDITIONS FOR SUCCESS

- Government resolve
- A sound political and social case for the selection of a permanent forest estate
- Long-term security for the forest estate, once chosen
- A market for forest produce
- Adequate information for the selection of the forest estate and for planning and controlling its management
- A flexible predictive system for planning and control
- The resources needed for control
- The will needed by all concerned for effective control.

NOTES

1. International Tropical Timber Agreement, 1983 (United Nations: New York, 1984). See also: H'pay, Terence, *The International Tropical Timber Agreement: Its Prospects for Tropical Timber Trade, Development and Forest Management*, IUCN/IIED Tropical Forest Policy Paper No. 3 (London, 1986).

2. The report deals with the management of closed tropical forest in the sense defined by FAO in its authoritative publication *Tropical Forest Resources* (Rome, FAO: 1982). Mangrove forests are not included.

3. FAO, *Management of the Tropical Moist Forests in Africa*, FAO forestry paper (Rome: FAO, in press); FAO, *Management Systems in the Tropical Mixed Forests of Asia: Studies in India, Malaysia and the Philippines*, FAO forestry paper (Rome: FAO, in press); Schmidt, R.C., *Tropical Rain Forest Management: a Status Report* (Rome: FAO, 1987).

4. HIID, *The Case for Multiple-use Management of Tropical Hardwood Forests*, Study prepared for ITTO by the Harvard Institute for International Development (Cambridge, Mass., 1988).

5. Hadley, Malcolm (ed.), "Rain forest regeneration and management. Report of workshop, Guri, Venezuela, 24-26 November 1986", Special Issue 18, *Biology International* (1988) (International Union of Biological Sciences).

6. Mergen, F. and J.R. Vincent, *Natural Management of Tropical Moist Forest* (Yale, 1987).

7. Repetto, R. and M. Gillis (eds), *Public Policy and the Misuse of Forest Resources* (Cambridge: Cambridge University Press, 1988).

8. Burns, D. *Runway and Treadmill Deforestation*, IUCN/IIED Tropical Forestry Paper No. 2 (London, 1986); IIED/Government of Indonesia, *Forest Policies in Indonesia: the Sustainable Development of Forest Lands* (London: International Institute for Environment and Development, 1985); Wyatt-Smith, John, *The Management of Tropical Moist Forest for the Sustained Production of Timber*, IUCN/IIED Tropical Forest Policy Paper No. 4 (London, 1987).

9. *Pre-project Report: Natural Forest Management for Sustainable Timber Production. Vol. V. Bibliography* (London: International Institute for Environment and Development).

Acknowledgements

This work has only been possible because of the help of a very large number of people in many countries. In particular, the directors of Forest Departments in all the producer countries of ITTO gave us the greatest possible co-operation and assistance, and made available to us the resources and experience of their staff.

My thanks go to the three regional consultants, Peter Burgess, Simon Rietbergen and Dr Timothy Synnott for covering a great deal of ground against strict deadlines; to John Palmer (of the Oxford Forestry Institute) for distilling his experience into his contribution to the study; to Dr Jeff Burley of the Oxford Forestry Institute for his help; and to John Wyatt-Smith, Dr Mike Ross and Ron Kemp for their wise comments; and to Terence H'pay and Alf Leslie for their encouragement.

The staff of ITTO have given constant help and encouragement. I should like especially to express my appreciation to Dato Dr Freezailah, the Executive Director, and to Nils Svanqvist.

We have also received warm co-operation from the Forestry Division of FAO. My thanks go to Dr Jean-Paul Lanly and to Ralph Schmidt for making available to us detailed material, often unpublished, from the files of FAO and giving us comments and advice from their wide experience.

The World Conservation Monitoring Centre has helped us with information on protected areas; our thanks to Dr Jeremy Harrison.

In the producer countries we were welcomed and received assistance and valuable information from very many individuals in headquarters, in the field and in the regional seminars. We thank the following:

Bolivia: Javier Lopez Soria; William Simons Rana; Juan Ramirez Pinto; Guillermo Lenz; Mario Saavedra Ramirez; Juan Carlos Quiroga Mendizabel; Dr Howard Clark; Raymond Victorini; Tage Michaelson; Gerardo Losano; Luis Goitea Arze; Esteban Cardona; Mariano Lozano; Edgar Landivar Landivar; Humberto Castedo Leigue; Federico Bascope; Lincoln Quevedo; Oscar Adan Marinez;

Dr John Wilkins; James Johnson; Williams C. Cabrera Colombo; Dr Ernst Schlieder.

Brazil: Emanuel Cruz; Olegario Pereiro de Carvalho; Jorge Alberto Gazel Yared; Osmar Jośe Romeiro de Aguiar; Silvio Brienza Jr; Jośe do Carmo Alves Lopes; Waldenei Travassos Queiroz; Juria Jankauskis; Fernando Antonio Bemergui; Sueo Numazawa; Manoel Sebastiano Pereira de Carvalho; Tania Linhares da Silva; Sergio C. Coutinho; Maria Joaquina Pires; Marcelo Vieira Albuquerque; Euclides Reckziegel; Joel Santos Gomes; Jośe Ribamar; Armando Pinheiro Carvalho; Cesar Agusto Carneiro Lopes; Luiz Manoel Pedroso; Virgilio Mauricio Viana; Rionaldo Rolo de Almeida; Prof. Dr Jurandyr da Cruz Alencar; Joaquim dos Santos; Gil Vieira; Alain Coic; Niro Higuchi; Harry van der Slooten; Gilberto do Carmo Lopes Siqueira; Jorge Ney Macedo Neves; Dr Carlos Marx Ribeiro Carneiro; Sergio Alberto de Oliveira Almeida; Sebastião Kengen; Paulo Lopes Viana; Paulo Galvao; Valdemar Carneiro Lēao; Sergio Barreiros de S. Azavedo; Jean Dubois.

Cameroon: Bénédict Fultang; Hubert Simo; Jean Makon Wehiong; Félix Essame; J.-B. Samgba Ahanda; Roland Camirand; Emmanuel Ze Meka; Roger Foteu Kameni; Adolphe Obam; Ferdinand Nkolo; Michel Cantin; René Dubuc; Jean-Baptiste Fotsing; Jacques Yougang; Jean Williams Sollo; Dieudonné Njib; David Momo; M. Chitou; Christine Chaperon.

Congo: Jacques Kanwe; Grégoire Nkeoua; Albert Essereke; Faustin Otouba; Pierre Sebgiolt; Ota Setzer; M. Moyo; V. Bouetoukadilamio; Dominique N'Sosso; Michael Askwith; Hubert de Foresta; Christian de Namur; J.C. Delwaulle; Philippe Vigneron; Eric Forni; Geneviève Michon.

Côte d'Ivoire: Denis Konan; Konan Soundele; P. Mengin-Lecreulx; M. Mallet; M. Thiel; Jean-Guy Bertault; Jozsef Micski; Fred Vooren; Aïdara Gouesse; M. Yamke; M. Kadja; Charles Kattie; M. Soro.

Ecuador: Mario Pescarolo; Fernando Escobar Suárez; Jośe Vicente Vallejo; Pablo Roser; Oswaldo Vivanco; Nelson Toledo; Mario Torres; Ing. Ponce; Ruth Quesada; L. Benitez; Nikolaus Henning; Juan Salinas; Vicente Molinos; Carlos Burbano; Alberto Robalino; Jorge Barba; Roque Miño; Roberto Pena; Fernando Montenegro Sanchez; Yolanda Kakabadse; Cecilia Pacheco.

Gabon: Jean-Boniface Memvie; Ondo Nkoulou; Gérard Dufoulon; Joseph Mbogho Oyame; Faustin Legault; Thomas Ekome Mbenge; Gabriel Azizet; Raphael Dipouma; Mathieu Ndong; Ard Louis; Laurent Rivière; Mme Drouilh; Gahuranyi Tanganika; Rigobert Ngouolali; Darius Posso Paul; Gérard Lematre; Denys Frere.

Ghana: Johnny François; Francis Atta Friar; Jean Djigui Keita; E.K. Afanyedey; Mr Amuzu; James K. Aggrey; Dominic D. Arhin; J.P. Peprah; A.S.K. Boachie-Dapaah; Dr Rex E.O. Chachu; Mr Gharty; Sam K. Riley-Poko; K.A. Sereboe; H.G.B. Smith; S.P.K. Britwum; Hugh Blackett; Bill Howard; Dr E.O. Asibey.

Honduras: Jośe Segovia Inestroza; Manuel Hernandez; Imerto Rosales; Jorge Palma; Jorge Fu; Luis F. Valle; T.W.W. Wood; John Roper; Julio Barahona; Mateo Molino; Rigoberto Romero Meza; Cesár Augusto Alvaradfo; Jośe Armando Valle; J.L. Montesinos; Oscar Ochoa; Salvador E. Romero; Anthony Wolffsohn; Thomas Hawkins.

Indonesia: Dr Saryono; David Boulter; Mr Djaret; Roesdin E. Akasse; Goenawan Wirisono.

Liberia: Benson Gwyan; Samuel G. Bobway; John W. Toe; Moses A.F. Carter Sr; Peter Weinstabel; Shad G. Kaydea; Emanuel Eme; James Mayers; Boakai Kawah; Eric Kamara; Albert B. Gbanya; Sormongar S. Zwuen; Randolph Miller; Mazein R. Fardoun; Emanuel Davis; Philip Jokolo; A. Momoh Kamara; Safu Gongar; Reinhard Wolf.

Malaysia, Peninsular: Dato Mohd Darus bin Haji Mahmud; Dr Salleh Mohd Nor; Mok Sian Tuan; Lal Singh Gill; Dr Thang Hooi Chiew; Dato Abdul Latif bin Nordin; Enche Ahmed bin Zainal; Tay Soon Toh; Enche Zul Mukshar bin Dato Mohd Shaari; Enche Shaharuddin bin Mohd Ismail; Enche Azmin bin Nordin; Mr Singham.

Malaysia, Sabah: Miller Munang; Daniel Khiong Kok Sin; Yahya Awang; Dr S.K. Bhargava; Herman Anjin; Vincent Fung; Ee Cheng Tak; Tang Hon Tat; Cyril Pinso; Dr Clive Marsh; P.E. Braganza; A.J. Hepburn; Roberto Darya; En Rashid; George Tan; Frederick Liew; Tan See Koong; Roger Onkili.

Malaysia, Sarawak: Lee Hua Seng; Ralf Ly Chong.

Papua-New Guinea: Andrew Tagamasau; Francis Cortes; Sisiric

Siniwin; Bernie Cuyne; Kamung Matrus; David Faunt; Philip Siaguru; Professor Robert Johns; Brian Kingston; Victor Buenaflor; Steve Layton; Gordon McNeil; Paul Unwin; Ken Hart.

Peru: Carlos Romero Pastor; Wilfredo Ojeda; Emilio Alvarez Romero; Rodolfo Taboada; Moises Trujillo; Guillermo Chota Valera; Jorge Malleux Orjeda; Raul Parraga Solis; Renato Ruiz Gutierrez; Denis Buteau; Carlos Llerena; Hanibal Chung; Roberto Lopez; Dr Daniel Marmillod; Pedro Reyes; Ing Koike; Teodoro Trucios; Enrique Quintana.

Philippines: Cirilo B. Serna; Bayani S. Nora; Rene Siapno; Roberto Malabanan; Dr Jurgen G. Schade; Dr Jochen B. Weingart; Jose Elmer C. Bascos; Jose Lechonoite; Ricardo G. Santiago; Roberto A. Dermonde; Dr Antonio Glori; Jose Kanapi Jr; Mr Agbayani; Durus E.B. de Guzman; Dr Delfin Ganapin Jr; Dr Oca Hamada; Dr Jim Samura; Dr Nellie Huertozado; Dr Rafael Ligayu; Jose S. Buenaobra; Emitiano Carpo; A. Willy Pollisco; Delfin Ganapin.

Thailand: Chumni Boonyobhas; Bhadharajaya Rajani; Surasak Ruangchan; Apiwat Sretarugsa; Vigoon Poo-Ri Tat; Sa-Nguan Kakhong; Dr Tim Smitinand; Sayan Jarukom; Choosak Chanprasith; Jiamsak Wichawutidong.

Trinidad and Tobago: Selwyn Dardaine; Clarence F. Bacchus; Sheriff Farzool; Carlton Sambury; Gerard McVorron; Robert K. Ramjohn; Thomas Gill; Sookram Ramroop; Boysil Rambharat; Eden A. Shand; J. Krebs; Rudolph Ramcharan; S. Ramsahai.

Throughout the study we have received great support from the staff of IIED. My thanks go in particular to Dr Caroline Sargent, Director of the Forestry and Land Use Programme, and to Dr Edward Barbier for their support and critical comment; to Elaine Morrison for much valuable editing; and to Carol Lambourne, Rachel Weinstein and Rashida Hemani, all of whom worked very hard to produce a presentable text of the report in time for the ITTO Council meeting and to prepare this book for publication.

Finally, my gratitude to my wife, Judy Poore, for her patience and critical proofreading.

The views expressed are those of the authors.

Thanks are also due to the Director of the Oxford Forestry Institute, through which we were able to secure the services of John Palmer and Timothy Synnott.

1. The Sustainable Management of Tropical Forest: the Issues
Duncan Poore

There can be few people now who are not aware that "the tropical forest" is in danger. This has become one of the clarion calls of environmentalists, a call which is now echoed in the statements of political leaders in both the developing and the developed world. It is one of the main issues identified in the report of the World Commission on Environment, the Brundtland Commission,[1] and there is no dearth of international action. But there is still much misunderstanding about the nature and scale of the problem and, above all, about what might be done about it.

This widespread concern about tropical forests is based on a number of issues: that these forests are disappearing at an alarming rate; that the loss of so much forest has potentially disastrous environmental effects – on soil, water, climate, the genetic richness of the globe and the supply of possible future economic products; that the uses to which the land is being converted are often not sustainable – that the forest in fact is being destroyed for no ultimate benefit, and that forest-dwelling peoples are being arbitrarily displaced. All of these are true – partly and in some places. And, as usual, it is drama and disaster that seize the headlines.

But there is another side of the coin. Successes can be chalked up here and there, examples of good planning and management, of conservation. They too make news – with the better correspondents. But they are still pitifully puny and infrequent in relation to the scale of the problem; and often, good though they may be, they are of only local relevance. The challenge is to extract from these local successes some general principles and to apply these widely and rapidly at a scale which will have some real impact.

First, though, it is necessary to find out what the real issues are, to separate what is important from the verbal froth. How vital really is it to preserve forest cover in the tropics? How much forest should be preserved intact and where? Is it possible to use some parts of the forest for economic purposes while maintaining environmental values? How might this be done? Where have planning and management failed? What are the reasons for failure and success? What can genuinely be done to solve these problems in the prevailing political and economic context of the tropical countries in which the forest is found?

This book does not set out to be a comprehensive account of tropical forest destruction; others aspire to do this. Instead, it concentrates on the management of tropical forest for the sustainable production of timber. It cannot do this, however, without trying to set this very important use, the production of timber, in a more general setting and examining some of the truths and myths of the tropical forest story.

It is justly pointed out that in the past much forest has been cleared elsewhere in the world, indeed that development has apparently often been based on forest clearance. Why should tropical countries be denied the same opportunity? But this clearance has been mostly in those climatic zones of the world in which agriculture has proved readily possible – the temperate regions, Mediterranean climates, the semi-arid zone, the subtropics; only in the boreal zone has most of the land surface remained under forest. Even in tropical and equatorial climates fertile alluvial soils were deforested centuries ago for irrigated agriculture. What then is so different about the remaining tropical forests?

Three points can be made here. First, it is now recognized that the deforestation in many of the long-established seats of agriculture was a very destructive process which led to much unnecessary loss of soil; if it were possible to develop those areas now with all the advantages of present knowledge, any wise government would hope to do it very differently. Second, there are good reasons why most of the remaining areas of tropical forest remain as forest: many, though not all of them, are on lands which have been inherently difficult to settle or cultivate, either because of disease or the infertility of the soil. And third, the forests of the wet tropics are exceptionally easy to destroy utterly, with all the plants and animals they contain. In most other parts of the world, forests can be destroyed but most of the tree species survive; and even many of the other woodland organisms can exist outside the forest or in small fragments of it. This does not seem to be the case in the wet tropics; so exceptional care is needed if it is wished to develop these lands in ways which do not waste their resources.

There is another consideration. Everyone is the product of the age to which he or she belongs. Developments which occurred in previous centuries took place in a climate of opinion and balance between population and potential resources which were very different from those of today. Liberties were taken with human rights and with the environment which would no longer be acceptable to world opinion. It is mistaken to suggest that these concerns are confined to the developed world, and that it is solely those from developed nations who are attempting to force their values on other less fortunate countries. It is evident that similar concerns are being voiced spontaneously by thinking people in the developing nations too and that these changes are signs of an inevitable progression in human conscience.

It is as a result of this new consciousness, and conscience, that the Tropical Forest Action Plan (TFAP) has come into being and that the International Tropical Timber Agreement (ITTA) contains the historic clause about "sustainable utilization and conservation of tropical forests".

What then is necessary for the sustainable utilization and conservation of tropical forest lands? In the following section of this chapter, we examine some of the concepts involved and try to set out the questions that need to be asked, and the standards that should be set, if sustainable utilization and conservation are to become realities in practice.

SOME DEFINITIONS

In order to make the study as exact and as useful as possible, some definitions are necessary; much confusion and misunderstanding can be, and have been, caused by the imprecise use of terms and we wish to avoid this as far as possible. Even in the subject of the study there are two terms that may be misinterpreted: "natural forest" and "management for sustainable timber production". Some explanation is given below of the sense in which certain key terms are used in this book.

Natural forest

This term is used in contrast to forest plantations ("forest crops raised artificially either by sowing or planting"), which are in general areas in which the naturally occurring tree species have been totally replaced by planted trees. Natural forest includes a range of types which have been subjected to varying degrees of modification by man and which grade almost imperceptibly into one another. The following list illustrates this range, starting with those that are least modified:

(a) virgin forest (compare with b, c and d below), essentially unmodified by human activity; this will contain gaps caused by the normal death and regeneration of trees and may include areas or phases which have been affected by natural events such as landslides, typhoons or volcanic activity
(b) forest, similar to the above, the composition and structure of which may have been modified by the hunting and gathering activities of indigenous peoples
(c) forests which have been subjected to a light cycle of shifting cultivation or in which cultivation has been abandoned, so that a full tree cover of indigenous species has been able to develop
(d) forests which have been subjected to various intensities and frequencies of logging, but which still remain covered with a tree or shrub cover of indigenous species. These may be of two kinds: those in which new tree growth is entirely derived from natural regeneration and others where this has been supplemented by "enrichment planting".

The first two of the above categories will be referred to as *primary forest*, the second two as *secondary forest*.

The following, both of which can make an important contribution to timber supplies, will be excluded from the definition of natural forest: areas which have been so intensively modified by cultivation, fire or other disturbance that they remain covered with grass or non-forest weeds – *degraded forest lands*; and forest plantations (as defined above), whether of native or introduced species.

Sustainable timber production

This phrase requires considerable explanation, both in itself and in relation to the more general term, "sustainable management". The relation between these two will be explained below.

When primary forest is first logged it normally contains a high standing volume of timber, a variable proportion of which is marketable, depending upon composition and market demand. Because this standing volume has accumulated over a long period, the commercial timber is likely to be of a quality and volume that will probably not be matched in future cuts (because it contains slow-growing specimens and species, large diameters, etc.) unless the logged forest is closed to further exploitation for a century or more. In this sense the first crop is, in practical terms, not repeatable.

If production of timber is to be genuinely sustainable, the single most important condition to be met is that nothing should be done that will *irreversibly reduce the potential of the forest to produce marketable timber* – that is, there should be no irreversible loss of soil, soil fertility or genetic potential in the marketable species. It does not necessarily mean that no more timber should be removed in a period of years than is produced by new growth; overcutting in one cycle can, at least in theory, be compensated by undercutting in the next or by prolonging the cutting cycle.

But even this description of sustainable production begs several questions. For example: markets will certainly change between one phase of logging and the next – new species will become marketable and current fashions may decline. So the "timber production" to be sustained will not be the same from one cutting cycle to the next; it will contain a different mix of species. One consequence of this is that valuable species may be over-exploited initially for economic reasons.

Moreover, the forest will certainly alter somewhat in composition as a result of selective harvesting and there may even be some loss of soil fertility.[2] But potentially damaging changes of this kind can be compensated by new investment – in these instances by the application of fertilizers or by enrichment planting. Whether to invest or not is essentially an economic decision.

The best silvicultural practice requires the calculation of the volume of timber which may be cut in one year in a given area (the "annual allowable cut" – AAC), a volume which should be set at a level that provides the maximum harvest while ensuring that no deterioration occurs in the prospects for future sustainable harvests.

When an area of virgin forest is cut the AAC depends upon the volume of marketable timber in the area which may be cut while leaving enough stems on the ground for the next crop. This is calculated from an inventory of the standing stock and an estimation of the length of the cutting cycle – the period between successive loggings. If the forest is well stocked, this figure may be high. But the situation alters for subsequent cuts, because the rates of growth of the remaining trees are changed when the first crop is removed. At this later stage it should be possible, by measurement of permanent sample plots, to determine a stable AAC which corresponds to the annual growth in volume of the forest in question. This is likely to be different from, and often but not always lower than, the AAC from the primary forest.

Sustainable production per unit area and sustainable supply

There is sometimes also confusion between these two terms because of the use in trade statistics of "production" in quite a different sense from that defined above – to mean, in effect, the supply of timber from whatever source. We have found representatives of some countries using "sustainable timber production" to mean continuity of supply from the natural forest, implying that when one source is exhausted, another will be found. It need hardly be remarked that this usage is dangerous, for it need not include any provision for continuity of production on sites exploited and can lead to the total destruction of the resource. In fact, in this sense, supply is sustainable until it runs out; then it is someone else's problem.

Management and its intensity

Management in the broadest context can be defined as taking a firm decision about the future of any area of forest, applying it, and monitoring the application. Dr Synnott, in his account in Chapter 4, writes:

> The term "management" is sometimes used loosely. Some activities are called "management projects" which actually consist of demonstrations and trials. Equally it is often said that management would be uneconomic when it is meant that silvicultural treatments would be. Here, management is understood to include many possible components: silviculture is one component of management, but it is possible to have effective management with few or no silvicultural interventions, and with only a few of the more important management activities.
>
> Here we distinguish between the characteristic tools of *silviculture* . . . regulation of shade and canopy opening, treatments to promote valued individuals and species and to reduce unwanted trees, climber cutting, "refining", poisoning, enrichment, selection . . . and *management* objectives, yield

control, protection, working plans, felling cycles, sample plots, logging concessions, roads, boundaries, prediction, costings, annual records, and the organization of silvicultural work.

The tropical forest can be managed for the sustainable production of timber at a number of different levels of intensity. This is often misunderstood and it is assumed that, if the forest is not being managed intensively, it is not being managed at all. We may take five different levels as examples, starting at the lowest:

Wait and see
Where forest is remote and there is, as yet, no market for logs, the most effective management may be to demarcate the forest and protect it from encroachment until it becomes worth while to extract timber.

Log and leave
In this "extensive" form of management, after logging, the forest is closed and protected from encroachment or further logging. The speed of recovery and the volume of the future crop will depend upon the kind of forest, the nature and standard of the first logging and the length of time that the forest remains closed.

Minimum intervention
Marked trees (up to the limit defined by the AAC) are removed with minimum damage to the remaining stand and according to a well-researched silvicultural system, leaving behind an adequate stocking for the next crop to be taken after a defined number of years. Removals are confined to stems intended for sale and defective stems of marketable species; no other species is interfered with. The area is then closed and protected without any further tending until the next logging is due.

Stand treatment
Logging is carried out in the same way as in the previous system, but the growth of the remaining stems or regeneration is enhanced by various treatments, which may include poisoning of unwanted stems or species, poisoning of lianes, weeding, etc.

Enrichment planting

The treatment is the same as "stand treatment" above, but saplings of desirable species are planted where the stocking of residual stems is low; special treatment is given to encourage these saplings.

All of these, if properly applied, would constitute management for sustainable timber production. Both the benefits, in terms of timber sales, and the costs, in terms of protection, tending, planting stock and chemicals, increase from the first option to the last. The decision about which to use at any time is essentially a matter of policy which is likely to be strongly influenced by prevailing costs and likely benefits.

In assessing the reports from the countries surveyed in this study, it is very important to recognize that these different levels of management may exist, at least potentially, and that all may be justifiably termed sustainable. On the other hand, any system, however good, may produce results that are unsustainable, if it is not applied conscientiously and consistently. Many good systems have failed because they are so complicated that there is neither the will nor the ability to operate them properly for even a few years, far less for as long as a scientifically based cutting cycle or even for the length of a rotation.

THE CONTEXT

The management of natural forest for the sustainable production of timber cannot be isolated from other aspects of national life, and these inevitably have an influence on the priority that is afforded to forest management in national policy and on the way it is regarded by various sectors of the government and of the people. This study has established that the success or failure of natural-forest management depends upon a great variety of different factors; but that technical constraints, although they certainly exist, are much less important than those that are political, economic and social.

Before entering upon a discussion of the findings of the country studies, there are a number of issues surrounding natural-forest management which need to be aired. These concern: the trade and the future sources of its timber; the national economic context of forest management; sustainability and environmental quality; and the social setting. These are far from being academic questions: they are all of great importance in considering whether natural-forest management and trade in tropical hardwood timber have a real future.

The trade and its future sources of timber

Tropical timber can come from a number of sources: (a) from previously unlogged forest which it is intended should remain as forest; (b) from logged forest which is being logged again, either with or without a plan of management; (c) from unlogged or logged forest which is being converted to another use in either a planned or an unplanned manner; (d) from secondary regrowth; (e) from trees planted in association with agriculture, roadside and canal-bank trees, etc.; and (f) from forest plantations.

Accurate statistics are lacking to determine the relative amounts which come from these different sources. There is little doubt, however, that most of the present supply to the market comes from the first cut of previously unlogged forest (a); substantial amounts come from both (b) and (c), though, as we shall see, very little of that from logged forest is under an operational plan of management; and small amounts so far from regrowth (d), agroforestry (e) and plantations (f).

There are many uncertainties about the future of these categories and the boundaries between them are by no means distinct. For example: (a) much forest that is planned to remain as forest does not remain so in fact; (b) little logged forest is under a plan of management which is effective and strictly applied; it is likely therefore to be deteriorating in quality and potential productivity; relogging is generally determined by the availability of a market rather than by considerations of sustainable silviculture; if light, it may be sustainable but, if not, it is likely to lead to progressive degradation of the resource; (c) clear felling, carried out when the forest is being converted to another use, results from a perfectly legitimate land-use decision, but permanently removes the forest from timber production; (d) there are no accurate trend statistics about what is happening to degraded forest or to sparsely populated or abandoned land; according to circumstances, it may degrade further or recover naturally and become capable of providing a supply of fast-growing, light-demanding timber species. Both (e) and (f) have potential, but at present make a very small contribution to tropical timber supply.

In any country, as the supply declines from the logging of land which is designed for conversion to agriculture and from the first cut of previously unlogged forest, the future must lie in four possible sources: managed natural forest; managed secondary regrowth; agroforestry; and plantations. Meanwhile the immediate market reaction to shortages or very steep price increases in timber from traditional sources is likely to be a movement away from the countries where supply has declined to those which have a largely untapped forest resource. Grainger, for example, has predicted a strong shift from South East Asia and Africa to South America.[3] The temptation for these newly favoured producing countries will be to follow the downhill path of resource mining which has been pursued by many of their predecessors, largely because they have a resource which they consider to be infinite.

As previously unlogged forest is logged and forest designed for agriculture is progressively cleared of timber, it is highly unlikely that the supply of tropical timber can be sustained at expected rates by plantations alone, however efficient. *Either* the management of natural forest for the sustainable production of timber must be practised widely over an extensive forest estate; *or* unprecedented steps must be taken to promote production in secondary growth, in plantations and in farm forestry. The decision about what should be the right mix is one for each individual producer country but the nature of these decisions is of considerable moment for the timber trade, for ITTO and for each individual producer country.

The national economic context of forest management

There seems to be general agreement that most governments seriously underestimate the *economic* worth of forests (both as productive resources and for the services that they provide) and that, on the other hand, they do not appreciate the cost of transforming the capital of natural forest into other forms of capital. There seems to be little doubt, at least in the view of one school of economists, that if the full non-market benefits of forests were to be taken into account, the "sustainable utilization and conservation" of natural forests would prove convincingly economic.[4]

Apart from these economic considerations, there are very great differences in the way in which forests are treated in *financial* terms, differences which make comparison difficult.

At one extreme – and this situation is relatively common in Asia – profits from the sale of timber are expected to generate large amounts of foreign exchange for investment in national infrastructure or other forms of profitable enterprise; this is in addition to covering the cost of the government forestry service and providing funds devoted, in theory at least, to reforestation and forest management. In Indonesia, for example, timber was in the 1970s the largest earner of foreign exchange and is now the third; and in Honduras the forestry service is a self-financing corporation which, in addition, makes a contribution to general revenue.[5]

At the other extreme, in Queensland, the exploitation of tropical forest provides profit for all involved in logging and the wood industries, but the cost of the forest service is borne from general revenue, justified, presumably, on the grounds of supply of a valuable product, the employment provided by the industry and the environmental benefits of retaining forest on the ground. In between there are a number of variants: separation of functions between forest service and forest enterprise; different methods of partition of royalties, with huge differences in the amounts charged; different ways of providing (or not providing) for the costs of forest management, etc.

Under these varying circumstances it is nearly impossible from accessible information to judge whether natural-forest management is financially viable (or indeed to assess what is meant by this term!).

Sustainability and the environmental context

The International Tropical Timber Agreement is concerned with the managing and harvesting of natural forest to meet the conditions of sustainability, and indeed this is one of its most important purposes. Something has been said above about the meaning of "sustainable timber production"; but this is in fact only a small part of the more general concept of environmental sustainability – an idea which is now being treated with the utmost seriousness in all informed international debate.[6] It cannot any longer be ignored in considering the use of tropical forests for timber production.

One expression of the conditions for an environmentally acceptable policy for tropical forest lands is that set out by IUCN.[7] This emphasizes that is essential, if the conditions of long-term sustainability are to be met, that forestry for timber production should be part of a land-use policy that makes full and proper provision for the conservation of biological diversity and for the protection of those forests which are critical for environmental stability (on erodible and infertile soils or where the removal of forest would have harmful effects on local climate or even global climate). Some further explanation of this ideal pattern of land use and management is included later in this chapter.

The sustainability of management of any area of forest can be assessed from two different points of view related respectively to the desired product and to the state of the forest: (a) to maintain the potential of the forest to provide a sustained yield of a product or products; and (b) to maintain the forest ecosystem in a certain desired condition. Consideration of these two will sometimes lead to the same result, but not always!

Both, however, depend upon further definition and clarification: in the first case it is necessary to answer the question "what product(s)?"; in the second, "what conditions?"

This point can perhaps best be illustrated by examples.

What product?

The most natural interpretation of this from the point of view of the timber trade is the sustained production of specified timber products. It should be possible to produce a silvicultural system to meet a certain mix of products. But if the aim is to maintain the potential to meet a market that may have changed during the period of the rotation, the system may have to be modified to cope with this uncertainty. There is also an important distinction between harvesting the mean annual increment of a certain chosen species and maintaining the potential of the forest to provide a sustained yield. The situation becomes even more complicated if the yield of other forest products is to be taken into account, rattan for example or game animals.

What condition?

Environmental arguments are likely to concentrate more on the condition of the forest than on its products. These arguments will hinge upon what condition it is desired to maintain, and therefore on the proper position of any particular forest in national policies for the allocation of forests to different uses. Let us take as examples two extreme cases.

At one extreme it may be considered sufficient to preserve the "forest condition" – the function of the forest in relation to soil conservation and water circulation. In this respect a well-planned rubber plantation with cover crops may be an acceptable substitute for natural forest; and any system of extraction and silvicultural management that provides adequate soil protection and continuous vegetation cover will equally prove acceptable.

At the other extreme, if the purpose of management is to maintain the forest in as near as possible to a virgin state for genetic conservation and scientific study, this condition is only sustainable by leaving the forest untouched.

Between these two extremes are all intermediate degrees. For example, if the successful establishment of a certain timber tree depends upon birds that are themselves dependent on a non-timber tree, the forest can only be maintained by natural regeneration if the second species of tree is permitted to remain; but it would also be maintained by the artificial planting of the first species.

Thus, if one is to be strictly accurate, sustainability can only be defined in relation to a specified set of products and a specified condition. It may, however, be possible to design a system that is an acceptable compromise between a number of objectives. For example a "minimum intervention" system of silviculture may provide a reasonable and sustainable harvest of timber and retain a very considerable degree of the biological richness of the forest; but the more intensively a forest is managed for timber production the more its composition is likely to be simplified and impoverished in variety of species and of genotypes. (Even a style of harvesting which is of very low intensity but highly selective – "high-grading" – is considered by many to cause genetic deterioration of the stock.) An ideal land-use policy would compensate for the intensification of timber production in some areas by more generous provision for species conservation in others.

This ambiguity in the definition of "sustainability" and the need for a nation to have forests which meet a number of different objectives emphasize: (a) the cardinal importance of careful land allocation in land-use policy; and (b) the need for an exact specification of management objectives in relation to each tract of forest.

Indeed it is questionable whether one universal definition of "sustainable management" is useful, because it will lend itself to different interpretations by different interests. To avoid misunderstanding it is essential to be very clear which kind of "sustainability" we are talking about in each instance.

There are also wider questions relating to sustainability which should not be ignored. It seems certain that the next century will see some degree of global warming. How this will affect the tropical moist forest is not clear. But it evident from the forest fires in Kalimantan, Sabah and Brazil that even small climatic fluctuations can make forests which have rarely, if ever, burnt before, much more vulnerable to fire. This should be taken into account in determining standards of management. Extensive deforestation also certainly has an effect on the carbon dioxide balance and, perhaps, on regional climates. But in this respect any continuous tree cover is likely to be at least nearly as beneficial as natural forest.

Another issue of great significance will be the selection and management of protected areas against a background of climatic change. The near certainty that this will take place should clearly have a considerable effect on the criteria used for selection, the choice of boundaries and the relation between the management of the protected area itself and that of the areas that surround it.

The social context

The regulation of forest management has, in many countries, been considered to be exclusively the province of government, with exploitation being carried out by government itself or by its licensed contractors. In carrying out these functions the customary rights of local peoples have been or have not been respected to very different degrees. In Papua-New Guinea, for example, the great majority of the forest is considered to belong to the local people; whereas in some other countries customary rights are largely ignored.

It is generally true, however, that the benefits from government-promoted forest exploitation have seldom accrued to those who live in or near the forest. For hunter-gatherers timber extraction may seriously damage their environment, though the light extraction in some African countries apparently may have the opposite effect. But it is often the case that farming peoples seldom see the advantages of retaining forest as forest rather than converting it to another – to them more lucrative – use. The sustainable management of government-run forests depends, therefore, upon the resolve of governments, not only to manage the forest consistently for long periods, but to protect it from encroachment by a largely unsympathetic populace.

But there are other possible models for sustainable management. This study has shown that there are small-scale local successes in forests which are managed by private commercial concerns and by local communities; there may be possibilities in extending the use of these models much more widely. What is crystal clear is that no system works unless there is long-term security for the managed forest and unless those concerned can see that the enterprise is profitable to them. It must also, of course, be profitable to the country.

THE SUSTAINABLE DEVELOPMENT AND CONSERVATION OF TROPICAL FOREST LANDS

What does it mean?

What exactly is meant by the sustainable development and conservation of tropical forest lands? This is a question that should be asked – and answered; for the answer is by no means self-evident, and different people put many different interpretations on it. But it is important to reach some level of agreement, for governments and organizations will otherwise work at cross-purposes, and the objective will be that much more difficult to accomplish.[8]

The easy way of answering the question is to quote the various axioms and goals of the World Conservation Strategy;[9] and this is certainly a good starting-point, for these are both valid and wise. The main axiom of the World Conservation Strategy – that conservation depends upon development, and that lasting development is impossible without conservation – is an attempt to distil in a few words the essence of a very complex set of ideas. Like all such axioms it can easily be twisted by the unscrupulous to their own ends. Conservation *can* be an obstacle to development; and development *can* be very destructive. The goal must be to bring about development, where development is needed, *with* conservation.

Take an extreme example. Two of the three objectives of the strategy are to maintain essential ecological processes and to preserve genetic resources. The best way of accomplishing these in the tropical forest would be to maintain the *whole* forest untouched and inviolate. Soils would remain intact; the rivers would be moderate in behaviour; the water clear; and the wildlife abundant and in natural balance. Technological man would have no place. For these two objectives, preservation is the best form of conservation; there can be little doubt of this.

But such a course of action is no longer possible anywhere. Hence the third objective of the strategy: to ensure the sustainable use of species and ecosystems. This immediately throws up the questions: how much should be used, how much preserved?

There are other crucial questions associated with the word "sustainable". Land and water can be used in many different ways that are ecologically sustainable. It is possible to exploit the natural forest itself more, or less, intensively. The forest can be replaced by an artificial ecosystem (agro-ecosystem) and this, too, can be exploited sustainably at different levels of intensity, according to the amounts of energy and skill that are invested in caring for it. The more capital-intensive forms of management usually also carry their own costs – in higher use of energy, increased risks of pollution and so on. We must use *all* these different ways of managing natural resources if even the basic needs of growing populations in tropical countries are to be satisfied; further intensification cannot be avoided if the quality of life of these peoples is to be raised. Tropical forest conservation must therefore be concerned, first and foremost, with finding an acceptable balance between these different intensities of use. How much land should be used in each of these ways? Are there limits beyond which it is unacceptable to go? These are questions that must be faced in any tropical forest policy. They are not easy questions.

It is perhaps easier to understand the problem – though no easier to solve it – if one appreciates that the three objectives of the World Conservation Strategy are different in kind. The maintenance of essential ecological processes and life support systems is concerned with keeping the whole machine running smoothly and preserving the natural capital of productive soil; the sustainable use of species and ecosystems, with maintaining the capital of natural systems that are cropped; and the preservation of genetic resources, with an insurance for the future. Choices must be made all the time about how much should be invested in each of these.

The only way of providing for conservation totally effectively and securely is by leaving natural ecosystems intact, and then only if the intact areas are very large. Once they are managed they are changed. The slightest manipulation alters the natural balance of populations and species; as intensity increases, so does the probability of irreversible change. Populations of plants and animals are affected first; then whole ecosystems go; then soil. So conservation is concerned not only with the balance between different kinds of use but also with the way in which one use is changed into another, and with the standards of management at all levels of intensity.

In the last analysis land, resources and people *will* reach a balance; sustainable use is the only ultimate possibility. They may reach this equilibrium in a planned, humane manner and in such a way that relatively rich resources remain at the disposal of humanity, or by a series of cataclysmic plunges that cause untold hardship and leave an impoverished world. The first is the only sensible course to follow.

The balance of use

Is it possible to say anything useful about balance between different uses – about how much of the natural ecosystems should be preserved, how much exploited in a sustainable way, and how much converted to other sustainable uses? Guidance is needed at both the national and the global levels.

The management of its natural resources is the responsibility of each nation. Each nation will look upon these resources in the total context of its economic and social development – the present relation between resources and population, the present standard of living and distribution of wealth, predictions about trends in all of these, and national social and economic objectives. Each nation needs to work out a balance for the conservation and use of its tropical forest lands that is acceptable in this context. There can therefore clearly be no one answer to the question of balance which is valid for all times and all nations; nor is there a single path to reach that balance.

Then there is the world perspective, for many national actions have world-wide effects. What is required is a set of national policies for the balanced use of tropical forest lands that fit together, like the pieces of a jigsaw, into a sensible world picture of tropical forest conservation.

Conservation: the moving target

If the resources of tropical forest lands are to be effectively and lastingly conserved, it will not be enough to design answers that only fit the circumstances of today. There will be great changes even in the next few decades, and more in the next century. While conservation is concerned with attaining stable balances, development is dedicated to change. Policies should be imaginative and designed to adapt to changing circumstances: in the balance between populations and resources, in economic well-being, in world energy policies, the balance of trade and in the attitude of people to environmental issues.

Many land-use policies are obsolete before they are implemented. Conservation policies for tropical forest lands must, therefore, look forward, be integrated as far as possible with policies for population and for all sectors of the economy, and aim to hit a moving target.

What is the scope of tropical forest conservation?

Tropical forests are very varied in character. In the wetter, equatorial and tropical regions there are true rain forests. These are very rich in species, frequently occur on poor soils and, if cleared or mismanaged, can prove very fragile. Once the forest cover is destroyed, very few of their species can survive. Destruction or damage is being caused by agricultural expansion, planned or unplanned, and by bad forest exploitation; fuelwood cutting is rarely a problem until the forest has already been reduced to remnants. As the climate becomes drier the rain forests give way to semi-evergreen forests, then deciduous forests which gradually become lower in stature as the rainfall decreases. These drier forests have been much more extensively modified than the rain forests – by agriculture, grazing and burning. They have been replaced over large areas by savannas, grasslands, degraded scrub or wasteland. In such areas, shortage of fuelwood is often a problem. The mountain forests near the Equator have characteristics that are not shared by those outside the tropics (strong daily fluctuations of climate but small seasonal differences); those in the subtropics have much in common with the forest of more temperate regions. In all these forests there are great variations due to biogeographic history, and local differences in climate, geology, land form and soil. They are very heterogeneous.

It is important to realize that there is such variety in forest type included within the term "tropical forest" and that the problems facing these forests differ widely from place to place. Most generalizations break down in face of this variety. For example, in the regions where there is exploitation of tropical timber there is rarely a fuelwood problem.

Neither can attention be confined to those areas that are covered with forest. In many places the forest has been degraded to grassland or scrub; some of these secondary communities have their own flora and fauna and are worth preserving because of this. Some of them could and should recover to forest or be used for the development of new tree plantations. The problems of deforestation often lie outside the forest, for example in policies concerned with the expansion of agriculture. Some countries, such as China, India and Peru, are only partially countries of tropical forest; what happens to the forest can depend on priorities determined outside the tropical region. The policies for tropical forest conservation are concerned therefore in the first instance with the forest but, beyond that, with the future of tropical forest lands and the various factors and pressures that influence the future of these lands.

What is the starting-point?

In a very few countries, such as Gabon, Brazil and Cameroon, where the population is low and much of the land is still covered in intact forest, it may still be possible to develop an ideal policy; but in most there is strong pressure to bring forest land into economic use or to clear it for agriculture. Where there is not yet pressure of population or of economic development on resources, the forest and its soils can and should be looked upon as an asset upon which to build future national economic development; where there is severe pressure, there may be little choice but to clear to produce more food and fuel. In extreme cases, and they are not uncommon, the resource of the once-forested land is so degraded and the pressures so great that a gigantic task of restoration is required.

While there is undoubtedly value in defining the ideal conservation policy and the limits of the unacceptable, in many countries things have deteriorated so far that the only politically possible course of action may be to reduce the amount of the unacceptable and edge gradually in the right direction; the ideal may always remain elusive.

The elements of the ideal

Suppose, then, that one were given a free hand to design the ideal policy for a country which had almost all its forest intact but where there was a need for economic development to meet the reasonable requirements of a gradually expanding population: what should be the main elements?

Conservative land husbandry should be looked upon as a spectrum from land that is totally protected to land that is used very intensively. In between there may be many shades, different degrees of exploitation or restraint from use, according to the values that it is thought important to preserve.

Protection

The natural forest is a bank of land and of resources, secure, and maintained at little or no cost. Until it is needed for other purposes, all land should be kept under forest. The burden of proof should be on those who wish to alter or remove the forest. If it is necessary to intensify land use, areas which are not under forest should be chosen first if such are available.

Certain areas should be permanently dedicated to protective uses. If these were to be lost the loss would be irreversible. Absolute protection is not always required, however, to maintain the value of these.

Living room for indigenous peoples

In many tropical forest areas there are indigenous populations who are in harmony with the ecosystems in which they live. Provided that this harmony is maintained and that these people wish to go on living in this way, the areas which they inhabit should be protected for their continued use.

Fragile soils

Measures for the conservation of all fragile soils, and the resulting protection of water resources, should take priority over any form of use or development which would cause erosion or loss of fertility.

This means the mapping of all areas with fragile soils in order that no use or development which would cause erosion or loss of fertility should be allowed in them. In very fragile areas there should be a permanent ban on exploitation. In others, no use or development should be permitted until it has been established that this could be carried out, and later managed, in such a way that no soil deterioration would take place.

Genetic resources

Special measures need to be taken to preserve the intra-specific variation of species of economic importance. These would include protecting some areas specially for this purpose and applying management constraints in chosen areas elsewhere.

Samples of ecosystems

These include measures for the preservation of representative samples of widespread ecosystems (climax, sub-climax and extra-zonal) and examples of exceptional ecosystems.

The full range of species
Many species will be protected by the full range of measures above. These should be supplemented to include: areas of special interest because of their exceptional richness or unusual character; the feeding, roosting and nesting sites of migratory birds; and sufficient territory to encompass full populations of wide-ranging mammals and birds.

Forests for sustainable production
A sufficiently large area should be allocated for the production of timber, wood and other forest produce to meet foreseeable domestic requirements for the next fifty years at least; further areas should be set aside to satisfy export markets if there is no shortage of forest in the country. If this area of forest is likely to prove insufficient to meet domestic needs when managed in an environmentally satisfactory manner, the shortfall should be made up by intensively managed, highly productive plantations (wood farming).

Land for food
Where not required for conservation purposes or for productive forests, the best soils should be held available for agriculture; but they should not be opened up until there is a real need. The first call on this new land for agriculture should be for those in greatest need, especially those who have no alternative to overusing and degrading the land on which they live.

If pressure for essential food or wood makes it necessary to turn forest on fragile or vulnerable land into farmland, this should be developed with combinations of food and tree crops that retain as far as possible the protective characteristics of the forest structure. Where land is allocated to agriculture or wood farming it should be made as productive as possible to help to reduce pressures of overuse on other areas.

Other environmental safeguards
In addition, there should be appropriate environmental safeguards covering any major developments, such as large reservoirs, irrigation systems, urban settlements, communications, to ensure that they are in harmony with this pattern of land use and with the essential requirements of conservation.

Economic and social considerations

The above are the main ecological elements of an *ideal* policy for the sustainable development of tropical forest lands. But ecological sustainability is only one facet of the problem. If solutions are to be lasting, they must be economically viable and socially acceptable. It is necessary to develop towards the ideal in steps that are practicable from the situation which exists today in many tropical forest countries. The development of a policy and the adoption of the measures necessary for tropical forest conservation depend upon the evolution of appropriate social and economic mechanisms. At the national level, they need to be politically acceptable and to fit readily into the framework of sensible economic policies. At the local level, patterns of land use need to be acceptable to the people they affect and not to intrude harshly into the harmonious development of local communities. They should be based on a broad understanding of the ecological circumstances and of the social setting.

There are many obstacles to this harmonious development. Central, perhaps, is a lack of understanding that the intact forest and its soils *are* a capital resource. This is reflected in the plundering of this capital and a failure to invest in the protection and maintenance of the resource. But there are many other obstacles: political pressures, social unrest, considerations of national security; failure to treat land-use questions in a manner that is integrated and socially sensitive; and a lack of trained staff and the knowledge that can be derived from well-conceived research.

As indicated above there are many different starting-points. Some fortunate countries have the resources to plan immediately for the ideal. Others may plan to power their economic development by depending on timber and on the cash crops that can be planted on good forest soils. They should be encouraged to examine how far these objectives are consistent with the ideal (and therefore how far they are sustainable) and to move towards an ideal pattern of land use. Where the pressures to meet everyday needs are urgent and immediate, it may be totally impracticable to think in terms of all elements of the ideal; in these cases soil and water conservation, and the growing of food, fodder and fuel, are bound to take precedence in domestic priorities over genetic conservation. Here international planning may bridge the gap: it may be possible to preserve the same resources somewhere else or, if this is not possible, to compensate for national hardship. There are cases too where the sustainable use of resources is likely to remain a pipe dream unless there is an investment of international resources.

Planning considerations

It has been stressed earlier that the conservation target is constantly moving. At present we tend to think in terms of a pattern of development that will retain a substantial proportion of the rural people in the countryside. Where this is the case, it will often be necessary to plan an attractive package of measures to combine conservation with development in an acceptable way: a protected area with production forest around it, both of which provide local employment; the development of agroforestry and the intensification of agriculture; some forest plantations and local industry. As well as developing thriving local communities, these will serve to buy time.

But there may be countries where development will bring about a rapid change in the pattern of rural settlement and in social priorities. This has already happened in some places. It is necessary for conservation policies to anticipate these changes and plan accordingly; for example, depopulation of rural villages can lead to a reduction in demand for wood and fodder in the countryside but a growth in interest in recreation and rural amenities from a more affluent urban class. The change can be very rapid.

International dovetailing

There is an important international element in policies for tropical forest conservation. The effective conservation of tropical genetic resources is a matter for international planning and co-operation, as regards the choice and safeguarding of protected areas. The same is true of measures to protect species. But this in no way lessens the duty of each nation to manage its own resources in a responsible way; indeed it increases the responsibility by adding an international dimension.

But good land-use planning and proper sustainable use can become very difficult unless there are international policies which encourage them, or at least do not positively discourage them. International markets and policies can very readily encourage unwise use of resources: the movement into cash crops rather than food crops; over-exploitation of tropical forests caused by incentives to satisfy foreign markets; unwise decisions about land use as a result of the instability of commodity markets.

A very important (and difficult) element in national policies in relation to tropical forest conservation is to strive for the harmonization of international policies, particularly on trade and aid, so that these provide a real incentive to the sustainable use and conservation of the resource. The Tropical Forestry Action Plan is working towards harmonization of aid while ITTO is concentrating on trade.

DEFINITION OF ISSUES IN THE ITTO STUDY

In order to define the status of the "sustainable utilization and conservation" of natural forests in the countries visited and, in particular, the status of management for the sustainable production of timber, we attempted, therefore, to find answers to these general questions:

(a) Over what areas is natural forest managed at an operational scale for the sustainable production of timber?
(b) Where such management has been undertaken successfully, what are the conditions that have made this possible?
(c) Where such management has not proved possible or has been attempted but failed, what have been the constraints that have made it difficult or impossible to apply? (The conditions for success and the constraints tend to be mirror images of one another.)

In analysing the situation some more detailed questions proved to be useful, all of which are relevant to successful management. These came under the following headings.

Policy
Is there a national land-use policy? Is there a national policy for the sustainable management of a permanent forest estate? If not, why?

Extent
What area of natural forest is managed for the sustainable production of timber?

Allocation

Is there a satisfactory system for choosing, demarcating and protecting those areas that will be used as production forest? If not, why?

Is there a satisfactory system for choosing, demarcating and protecting those areas that will be used as protection/conservation forest? If not, why?

Are there pressures from other sectors or interests to remove productive forest from use? What measures are being taken to counter or divert these pressures?

Sociological and economic conditions

In what ways do the various people who have an interest in or are affected by the management of the forest benefit from this management or suffer from mismanagement (people dwelling in or near the forest, loggers, middlemen, wood processors, small industries, consumers generally, the Forest Authority, and other parts of government)? Are the benefits adequate to provide an incentive to good management? Is there equitable distribution of these benefits? If not, why?

Management

Are the objectives of management conducive to sustainable production? Are the management prescriptions appropriate for the particular forest type? Are they rigorously applied and reviewed? If not, why?

Pre-exploitation survey

How comprehensive and adequate is the pre-exploitation survey: choice and marking of trees for felling; analysis of trees to remain unfelled; existing regeneration; environmental conditions; routing of extraction roads? If inadequate, why is this so?

Choice of exploiters

Does the choice take into account the best long-term interests of the forest? How?

Conditions of exploitation

Do these bring reasonable benefits to the various parties concerned: government revenues, any reforestation fund, the logging companies, local contractors, logging labour, those with customary rights in the land?

Are the conditions of exploitation such as to encourage long-term investment in the sustainable management of the forest? Are there reasonable incentives to encourage good management? What proportion of revenues are returned to forest management? If the conditions are unsatisfactory, what prevents their improvement?

Quality of exploitation
Are there guidelines on the siting, construction and maintenance of extraction roads, on weather conditions in which exploitation should not take place, on the equipment to be used, on directional felling, cutting of lianes, etc.? Are such guidelines followed? If not, why?
Are the above conditions monitored during and after exploitation? How? How well?

Post-exploitation survey and treatment
Are there guidelines? Are they sensitive to different forest types? Are they adhered to? Is later performance monitored? How? If not, why?

Control and follow-up
Is there effective control of operations at all stages? If not, why? Are there arrangements for monitoring and reviewing prescriptions? If not, why?

Research
Is research designed to support sustainable timber production from natural forest? Does it provide the necessary information to answer the questions set out above? Are there permanent sample plots to provide the data upon which sustainable yield can be calculated? Are the data processed and made available to management within a reasonable time?

Education and training
Are enough trained staff at all levels being produced with qualifications in the skills needed in natural-forest management?

The answers to these questions form the basis of the next four chapters on the situation in Queensland, Australia, in Africa, in South America and the Caribbean, and in Asia.

NOTES

1. World Commission on Environment and Development, *Our Common Future* (Oxford, 1987).
2. John Palmer, in Chapter 6 of this book, expresses doubt whether the depletion of nutrients is ever a significant factor.
3. Grainger, Alan, *Tropform: a Model of Future Tropical Timber Hardwood Supplies, Proceedings of the CINTRAFOR Symposium in Forest Sector and Trade Models* (Seattle: University of Washington, 1987).
4. Burns, D., *Runway and Treadmill Deforestation*, IUCN/IIED Tropical Forestry Paper No. 2 (London, 1986); HIID, *The Case for Multiple Use Management of Tropical Hardwood Forests*, Study prepared for ITTO by the Harvard Institute for International Development (Cambridge, Mass., 1988); Repetto, R. and M. Gillis (eds), *Public Policy and the Misuse of Forest Resources* (Cambridge: Cambridge University Press, 1988).
5. For further figures see HIID, op. cit., Figure 2.2.
6. See World Commission on Environment and Development, op. cit.
7. Poore, Duncan and Jeffrey Sayer, *The Management of Tropical Moist Forest Lands: Ecological Guidelines* (Gland, Switzerland: International Union for Conservation of Nature and Natural Resources, 1987).
8. This section is based on a text prepared for IUCN and is reproduced with acknowledgement to IUCN.
9. IUCN, *World Conservation Strategy* (Gland, Switzerland: International Union for Conservation of Nature and Natural Resources, 1980).

2. Queensland, Australia: an Approach to Successful Sustainable Management
Duncan Poore

Australia is a member of the ITTO, though as a consumer rather than a producer nation. There are, however, considerable areas of tropical rain forest in Queensland which have produced valuable tropical timber for many years. Queensland was included in our study as an extension of the examination of sustainable management in the producer nations.

In many ways the situation there was found to approach the ideal postulated in the last chapter more closely than that in any other country visited. For this reason an account of the management of the Queensland rain forests is included here to illustrate the conditions under which one country has been able to achieve the sustainable production of timber from its natural forests and to provide a basis of comparison with the situation in other countries which have not so far been so successful.[1]

THE RAIN FORESTS

Distribution and character

The rain forests of North Queensland lie in a discontinuous belt some 400 km long and up to 50 km wide along the east coast from Mt Spec near Townsville in the south (latitude 19° S) to just beyond the Bloomfield River (latitude 16° S) in the north. This forest lies near the limits of a rainforest climate with an annual temperature range as high as 20°C near the coast. Cyclones are frequent.

The forests mostly occur on upland plateaux lying between 600 and 900 metres (with mountain peaks rising to 1,600 metres) connected to coastal lowlands by steep escarpments. Much of the region receives between 2,000 and 2,500 mm of rainfall with a marked wet season during the summer months, but there are areas which have as little as 1,250 or as much as 3,500 mm.

Towards the centre of its range the rain forest occurs as a relatively continuous block but it becomes progressively more interspersed with wet and dry sclerophyll (rigid-leaved) forest towards its north and south ends. There is evidence that its area has expanded and contracted during the last 120,000 years and that in the last 8,000 years it has been spreading from refuges in which it survived a period in which the climate was much drier than that of today. It is at present invading adjoining areas of wet sclerophyll forest which are protected from fire.

The rain forests occur on alluvial deposits, basic volcanics (basalts), acid volcanics (rhyolites), coarse granites, sedimentary and metamorphic rocks, and "Tully" granites. The most productive forests are on soils derived from the basic volcanics and granites.

There are approximately 800 tree species in the rain forest of which about 600 reach sawlog size; about 150 of these are harvested. (In 1981 35 per cent of the timber was used for veneer and 65 per cent for building. No timber is exported.)

Regeneration is abundant from seed; many species coppice; and some can spread by root suckers. In general the forest species are aggressive, as evidenced by the invasion of wet sclerophyll forest, and well adapted to disturbance. In this they differ from the rain-forest dominants in other parts of the tropics.

Policy

Forest policy is administered by the Department of Forestry under the provisions of the Forestry Act, 1959-81. Section 11 of the Act specifies, among the functions of the department: (a) the carrying out of all matters of forest policy; and (b) with the object of maintaining as far as practicable adequate supplies in perpetuity of timber and other forest products, the management of all state forests.

Section 33 defines as the cardinal principle of management of state forests the permanent reservation and management of these areas for the production of timber and associated products in perpetuity, having due regard to the conservation of soil and the environment, and also to other forest values including watershed protection and the use of the forest for other purposes such as recreation and/or grazing.

National parks are administered by the National Parks and Wild-life Service. Policy is to select representative areas of all forest types. In the rain forest thirteen types have been identified on the basis of structure. Where these are not adequately represented in national parks, samples of them are preserved in state forests. In addition, by means of surveys based on other criteria (bioclimatic, refuge status, containing rare species or associations, etc.), types of habitat have been identified that are still underrepresented, or not represented at all, in the present protected areas. Efforts are being made to remedy these deficiencies.

Table 2.1: Total area of rain forest in the wet tropics region of North Queensland, north of latitude 20° 30′ S

	VT (ha)	VF (ha)	Total by tenure (ha)
State forest	800	396,000	396,800
Timber reserve	–	108,500	108,500
National park	38,200	162,100	200,300
Other Crown land	64,100	110,100	174,200
Aboriginal com. areas	4,600	78,700	83,300
Private land	6,000	59,500	65,500
Total by type	**113,700**	**914,900**	**1,028,600**

Notes
Types:
VT   Vine thickets
VF   Vine forest (includes some areas of vine-fern thickets on moist windswept uplands in Townsville-Daintree-Cooktown and Cooktown-Cape York regions).

Tenure:
Other Crown land
    Crown leases on which forest products are available for harvesting only by the Crown; vacant Crown land; other reserves; dedicated roads.
Aboriginal community areas
    Land allocated for the exclusive use of Aboriginal inhabitants of Queensland. Land is held in various titles.
Private Land
    Freeholdings; Crown leases on which forest products are not available for harvesting by the Crown; land acquired by the Commonwealth government.

Source: Queensland Department of Forestry.

Area of the forests and status

It is estimated that the original rain-forest area of North Queensland amounted to about 1,200,000 ha; clearing for pasture and crops has reduced this to the present area of 1,028,600 ha. As a result, little rain forest remains on lowland alluvial soils.

In terms of recent management, the rain forest in North Queensland can be considered in five broad regions: Charters-Towers, Townsville-Daintree, Daintree-Cooktown, Cooktown-Cape York and the Offshore Islands. Estimates of the status of the rain forest are set out by type and region in Tables 2.1 and 2.2.

Out of the total of 1,028,600 ha, 161,275 ha (slightly over 15 per cent) are available for logging. Of this only a small proportion is virgin. In an assessment in 1981, the total net productive area south of Daintree was considered to be 142,749 ha, of which 22,534 were virgin, 44,869 logged before 1959, 37,364 logged between 1960 and 1969, and 37,982 logged after 1970.

Table 2.2: Total area of North Queensland rain forest by regions

	Charters-Towers (ha)	Townsville-Daintree (ha)	Daintree-Cooktown (ha)	Cooktown-Cape York (ha)	Offshore Islands (ha)	Total (ha)
State forest	–	396,700	–	100	–	396,800
Timber reserve	–	8,100	53,700	46,700	–	108,500
National park	28,600	108,400	18,500	38,000	6,800	200,300
Other Crown land	23,300	74,400	5,700	70,100	700	174,200
Aboriginal com. areas	–	7,300	–	73,100	2,900	83,300
Private land	5,200	51,800	6,900	1,500	100	65,500
Total	57,100	646,700	84,800	229,500	10,500	1,028,600

Source: Queensland Department of Forestry.

FOREST MANAGEMENT

Objectives

The management objectives for the North Queensland rain forests state that: ". . . areas available for logging are to be managed under multiple use principles with logging to be by a conservative selection system, with natural regeneration and subject to close control to minimise environmental impact. . ."

This policy has a number of elements, one set concerned with the continuing productivity of the forest, the second with other environmental values. Among the former are the determination of the annual allowable cut; keeping logging damage down to an acceptable level; the maintenance of the genetic quality of the growing stock; and concern about possible deterioration of the soil. Among the latter are the needs to safeguard the quality of water flowing from logged catchments; to reduce the amount of sediment in the water; and to manage the forest in such a way that species of plants and animals are not lost or, if they do temporarily disappear, can re-enter the logged area from adjoining forest. If the forest is to be managed in a way that accomplishes all of these objectives, there must be careful control of forest operations and a significant effort in research and monitoring.

This policy recognizes the environmental values of the forest, the economic value of the industry (about $Australian 30 million per year) and the direct and indirect employment generated by the industry amounting to about 2,000 jobs.

Timber production

Before 1978 sawmill allocations were set at 195,000 cu. metres, a level deliberately higher than the ultimate sustainable yield from the regrowth forests. This was because inquiries carried out in 1948, formally instituted by the Queensland government, found that there were adequate supplies of old-growth timber to maintain the high level of harvesting for many years. (Yields at first cut are of the order of 35-50 cu. metres per hectare, whereas the sustainable yield from subsequent management will be some 12-20 cu. metres per hectare.) Between 1978 and 1986 the allocation has been progressively reduced. Calculations indicate that the long-term average yield is in the vicinity of 63,000 cu. metres (corresponding to a growth of rather less than 0.5 cu. metres per ha per year) and the allowable cut for the period 1986-91 has been set at 60,000 cu. metres. New calculations will be made before 1991 to set the cut for the five following years.

The planned reduction in logging would have led to a regional net sawlog deficit from native forests by the year 2010 of 310,000 cu. metres, which it was planned to make up by the development of plantations and continuing importation of timber products into the region.

A new situation has arisen, however, since December 1987, when federal legislation was passed stopping all logging in those forests included in the site nominated for World Heritage status – in effect all the forests previously allocated for timber production. More recently the Queensland rain forests have become a World Heritage Site under the World Heritage Convention.

Silvicultural system

The forest is being managed according to a conservative selection system with a cutting cycle averaging about forty years. This selection logging aims to stimulate growth increment and regeneration in the rain forest through the removal of large-sized and defective stems. It favours more desirable species for retention as seed stock, especially the high-value veneer and fine cabinet woods.

This choice of system is based upon the measurements of thirty-seven growth and yield plots located in all the main types of forest, both logged and unlogged. These have been established for periods of from ten to thirty-five years and are designed to measure the performance of forest which has received no special silvicultural treatment. (They contain 399 of the 800 species of Queensland rain-forest tree.)

In addition, between 1945 and the early 1970s a number of experiments were conducted of post-logging treatment and enrichment planting, and twenty-seven research plots are maintained in these treated forests. Some experiments showed considerably enhanced growth of valuable species (especially enrichment planting with *Flindersia brayleana*) and, as a result, about 5,000 ha of forest were treated.

Following generally poor results achieved with plantations of rainforest hardwoods, plantings in the 1930s concentrated on the native conifer, hoop pine (*Araucaria cunninghamii*), on rain-forest sites. Planting on cleared forest sites ended at the start of the Second World War when there was a sharp increase in the number of species utilized. At this stage it was recognized that clearing of the rain forest destroyed much potentially valuable advance growth, some of which was on the verge of becoming marketable.

So far, plantations of rain-forest hardwoods have continued to prove unrewarding, the product being generally inferior to that coming from natural forest.

Calculation of the annual allowable cut

The annual allowable cut is based on a computer simulation technique known as cutting cycle analysis. The base data are: (a) stand tables derived from two-stage field sampling of 0.5 ha plots placed by air photo-interpretation and the national grid, and stratified according to the "loggability" of the forest; and (b) growth rates derived from research plots in virgin and logged forest. Refinements in technique have now made it possible to derive a more accurate figure from simulated harvests applied to individual plots.

In addition, this technique provides the basis for more detailed management decisions on the exact length of the cutting cycle, areas to be scheduled for logging, diameter cutting limits, the composition of the stand to be left as residuals and as seed trees, etc.

RAIN-FOREST LOGGING

Logging is carried out by licensees under agreements which cover the area of sale, the volume of timber to be extracted, and the conditions of operation of the sale.

Volume of timber

Detailed stand data are collected before harvesting begins; timber harvesting and the growth of the forest are both simulated; the yield of timber at various times in the future is predicted, and the capability for sustained yield of each stand is assessed. Timber marking is based on this assessment.

The planning of logging

There is a detailed sequence of planning and controls in the logging of any area of rain forest. It is as follows: inventory and yield calculation to establish the allowable cut; scheduling of areas from logging history records; development and maintenance of primary access; field investigation of net loggable area (locality, slopes, catchment values, special management values, inventory and stand simulation); log planning, in field design of roads, creek crossings, ramps, buffer zones; mapping the outline of the logging plan; offer of "sale" area to mills; construction of "sale access"; tree marking (see below); cutting and tracking; crowning; snigging (the dragging of the log from the stump) and haulage; the assessment of damage; provision for drainage of roads, snig tracks and ramps; sale closure for forty years plus; and the updating of logging history.

Tree marking

Tree marking is the means by which harvesting is regulated within the defined objectives of management. It is carried out by the Department of Forestry. The objectives of tree marking are:

(1) to harvest from the accumulated capital growth component of the forest while retaining a forest structure which is consistent

with the objectives of sustained-yield management. This implies the harvesting of all stems of merchantable species that are not required to ensure future harvests or for environmental purposes or genetic improvement

(2) to comply with defined environmental guidelines (see below)

(3) to encourage regeneration in accordance with the principles of sustained-yield management: this implies retention of an adequate seed source, and there are specific guidelines concerned with the retention of seed trees

(4) to manage species composition to improve growth and wood quality. This involves a number of tree-marking considerations: (a) species are grouped in terms of wood quality; Group A should be retained in preference to the other three groups; (b) cutting limit diameters for different species are set according to their growth potential; (c) the direction of fall is specified for each stem to minimize damage to growing stock; (d) stems to be retained which are very valuable or liable to logging damage (e.g. seed trees or those near snig tracks) are clearly indicated.

The spirit and intentions of the tree-marking guidelines should always be respected. They may be varied in a commonsense manner, for example in dealing with dangerous or faulty trees.

In general no trees are marked for felling except those of merchantable species, and guidelines should be modified to ensure that at least 50 per cent canopy cover remains over the productive area of the sale (excluding roads and snig tracks). Logging generally removes 15-20 per cent of the canopy; the maximum observed on the Windsor Tableland before the introduction of the environmental guidelines was 45 per cent.

The conservation of special genetic types, provenances or species is normally a function of the Scientific Area System but, where this is impracticable, the sales agreement may be modified to ensure their retention.

Certain species are protected from all logging except salvage logging (i.e. where land is required for other purposes), for example *Macadamia* sp.

Environmental guidelines

These are included in the logging plan and the map that accompanies it. The present guidelines came into force in 1982. Among them are the following, which give an indication of the level of detail prescribed; new regulations are even more detailed.

Sale area boundary. This is to be marked in the field; it is to follow creeks where possible and minimize the number of creek crossings.

Special management zones. These are to include areas of scientific, recreational and landscape significance; also erosion-prone or particularly steep sites where there should be no logging or logging only under special conditions.

Designated streams. These are all permanent or semi-permanent streams which, for reasons of erosion risk, wildlife habitat or recreational value, should remain undisturbed.

Haulage roads. These should be marked on the map accompanying the logging plan.

Major arterial snig tracks (the major common route to a ramp). Soil displacement (side cutting) should be avoided as far as possible.

Landings. Carefully located and as small as possible, their maximum size should be specified in the logging plan. They should generally be no larger than 750 sq. metres and with adequate drainage.

Roads. Clearing width for major extraction roads should be no more than 7.5 metres and for minor extraction roads no more than 5 metres. They should be located for ease of drainage and minimal earthworks, for example on ridge tops or moderate side slopes. Grades should generally be no more than 14 per cent.

Drainage. Regular drainage is essential for maintenance of all roads and tracks.

Filter/buffer strips. Required for the protection of designated streams. These are to be identified in the field and in the logging plan; they are to remain totally undisturbed except when approval is given in writing. The width of the buffer strip is specified according to the width of the stream, and there are also specifications of the size of catchment on different rocks over which the retention of strips is obligatory.

Temporary crossings. These should be located on sites with stable stream bed material and where bank restoration will be possible. If necessary the crossing should be corduroyed or constructed with stable gravel material.

Harvesting equipment. The type of equipment is to be specified in the logging plan. The blade width of any snigging machine should not exceed 4 metres and it should be fitted with a winch carrying not less than 30 metres of wire rope.

Weather and logging period. There should be no felling of trees or snigging or hauling of logs between 1 January and 31 March. Any exception to this must be authorized in writing and is limited to certain specified conditions. In addition, at any time of year, "the District Forest Officer may suspend all or any logging operations at any and for such time as he considers these operations are not in accordance with good forest management and practice."

Once logging is completed the snig tracks and landings are inspected to ensure that drainage is satisfactory, and the whole sale area is then closed.

Effects of logging

Measurements of damage in "salvage logging" (all trees over 40 cm diameter breast height – dbh) showed that approximately 20 per cent of the ground had been disturbed by snig tracks; but in normal logging the proportion is less. With the operation of the guidelines in recent sales, snig track disturbance is now down to 6–15 per cent. Also, as a result of the environmental guidelines, regeneration on disturbed areas is good: there is usually complete cover within five or six years and a mature forest structure has been restored before the next cutting cycle. The amount of damage to residuals marked for retention has been about 12 per cent since the introduction of the guidelines; loggers are penalized for such damage.

Most sedimentation has been found to originate from a few localized sources and can largely be stopped by the specified design of roads, tracks and creek crossings.

Financial considerations

The expense of rain-forest management falls mainly on the Department of Forestry, and of research on the Commonwealth Scientific and Industrial Research Organization (CSIRO). No calculations are available about how much this costs per hectare. Logging is profitable with sales at A$45–120 per cubic metre, stumpage at A$20 and costs of the order of A$40 (high-quality veneer timber can sell in Sydney for A$4,000).

RESEARCH

The silviculture and management of the Queensland rain forests depend upon a great volume of research carried out over many decades. Upon this in turn depends the maintenance of practices to harvest the forest in a sustainable manner and one which is environmentally acceptable. Some of this is time-consuming and costly; for example the measurement of research plots involves about 250,000 tree measurements each year. Present policies are based upon the results of this research but the situation is constantly being monitored to ensure that practices correspond to the objectives of management.

Research in silviculture has covered: (a) rain-forest treatment and natural regeneration; (b) growth and yield plots in logged and virgin forest; (c) underplanting or enrichment planting; and (d) plantation establishment.

In addition there has been much research into the environmental effects of silviculture and logging. This has concerned: the assessment of the degree of damage and disturbance in heavily logged catchments; erosion and sedimentation studies; interaction between water, the rain forest and its soils; and hydraulic studies.

There have also been studies on the effects of logging on flora, fauna and forest structure, their likely recovery times and measures to reduce adverse effects; and an assessment of the likelihood of ecological or genetic resource loss.

All of the results are taken into account in defining best practices for management.

CONCLUSION

It will be seen that the recent management of the Queensland rain forests and the pattern of land use in the rain-forest area approaches the ideal set out in the first chapter. No doubt, if planning were to start now with virgin forest, more provision would be made for nature conservation in the flat lowland coastal areas; and more attention would be paid to conforming from the very beginning of logging to a properly defined annual allowable cut. But on the whole the conformity is fairly good. The conditions for this have been: security of the forest estate; good control of all operations; profitability for the loggers and timber merchants; and adequate research to define a reasonable pattern of land allocation and the determination of the annual allowable cut.

Compared with most of the producer nations, however, the conditions in Queensland are unusual. In particular, there is no problem of agricultural encroachment; the forest department is centrally financed and the profits of exploitation are not required either to finance the department or to earn foreign exchange; and there is a strong tradition of well-funded research. On the other hand, environmental awareness has come relatively recently to Queensland. Perhaps as a consequence of this, the recent expression of the environmental conscience has been very strong and has led to the cessation of all logging in these forests.

NOTES

1. Anon., *Strategic Land Resource Zoning for the Wet Tropics Rainforest Region of Queensland*, A report by the Scientific Committee, Northern Rainforest Management Agency, Cairns, 1988; Department of Forestry, Queensland, *Timber Production from North Queensland's Rainforests*, Position paper (Queensland: Department of Forestry, 1981): Department of Forestry, Queensland, *Rainforest Research in North Queensland*, Position paper, (Queensland: Department of Forestry, 1983); Department of Forestry, Queensland, Leaflet series; Preston, R.A. and J.K. Vanclay, *Calculation of Timber Yields from North Queensland Rainforests* (Queensland: Department of Forestry, 1987); Vanclay, Jerome K., "A stand growth model for yield regulation in north Queensland rainforests" in Alan R. Ek, Stephen R. Shifley and Thomas E. Burk (eds), *Proceedings of IUFRO Conference on Growth Modelling and Prediction*, vol. 2 of Society of American Foresters Publication no. SAF-87.12 (St Paul, Minnesota: USDA Forest Service North Central Experiment Station, 1988). The IUFRO Conference took place 23-7 August 1987 in Minneapolis, Minnesota.

3. Africa
Simon Rietbergen

The importance of Africa's tropical moist forest (TMF) is huge. It provides environmental benefits such as soil and water conservation and stabilization of micro-climate and, reportedly, even of macro-climate. Its produce is important in financial terms, the value of the hardwood exports of the six countries studied in this chapter (Cameroon, Congo, Côte d'Ivoire, Gabon, Ghana and Liberia) amounting to US$500 million-600 million annually in 1985-7; it also gives a livelihood to millions of people who depend on the myriad non-industrial products it provides. Moreover, its biological richness is immense.

TMF includes wet evergreen, moist semi-deciduous, moist deciduous, and freshwater swamp forest. In Africa it is found from Sierra Leone to Ghana, and from Nigeria to Central Africa, stretching eastwards to Uganda and Tanzania, and southwards to Zaïre. A number of African TMF countries, such as Zaïre, the Central African Republic (CAR), Nigeria, Sierra Leone, Guinea and Uganda, have not, or not yet, signed the ITTA, and do not figure extensively in this chapter. However, where relevant, reference is made to these countries in more general terms.

This study concentrates on forest management for the production of industrial timber, but this actually forms only a very limited part of the produce and benefits provided by TMF in Africa. Indeed, in all countries studied, the most important contributions of TMF to the national economy, in both physical and economic terms (expressed mainly in replacement values, as most forest produce does not enter national accounts), would seem to be the non-industrial goods and benefits derived from trees or forests. These include the best upland soils, a variety of foods including bushmeat (often accounting for a major share of the animal protein intake of the rural population), fuelwood and charcoal, framing, panelling and thatching materials for rural houses, agricultural and household implements, and a host of environmental and other benefits.[1]

Management for sustained timber production, however, can be

wholly or largely compatible with most of these other values: with soil and water conservation, with production of most non-timber forest products, and even to a certain extent with nature conservation, depending on the intensity of management practised.

Why is natural-forest management for sustained timber production considered to be so important? After all, hardly any tropical timber is supplied by managed natural forest in Africa at the moment. Rather, the timber derives from forests which are being logged without a management plan, or are being converted to another use in either a planned or an unplanned manner.

However, these sources are expected to decline dramatically. In the West African countries studied all or nearly all of the TMF has already been logged over at least once. In the other African countries studied there are still large areas of primary forest in remote areas but accessible forests have been cut over several times. As timber exploitation in many of the remote forest areas is not economically feasible in the short to medium term because of the huge infrastructural investment needed, tropical timber supplies for some time to come will have to be derived largely from managed natural forest, supplemented by timber from managed secondary regrowth, agroforestry and plantations.[2]

In some scenarios, plantations are envisaged as the main source of tropical timber in the not too distant future. Indeed, plantations have long been seen as an alternative to natural forest management. From 1960 to 1980, almost all regeneration methods were artificial and the various management systems practised previously, such as the Tropical Shelterwood System, stand improvement methods such as "Amélioration des Peuplements Naturels" and various selection systems, were almost all abandoned.

But the reasoning behind this shift is flawed in two ways. First, economic expectations from plantations have often been unrealistic. Second and more fundamental, plantations cannot be a complete alternative to forest management, for they do not yield the same environmental and conservation benefits or the same variety of non-timber forest products as natural forests, and the timber grown is often inferior in structural qualities because of its faster growth.

Plantations have performed quite well as a source of industrial timber in a number of cases. With the best African timber species on good sites, a mean annual increment of 8 cu. metres per hectare per year may be expected at the age of thirty to forty-five years; but the high establishment cost of US$1,000-3,000 per hectare renders the profitability of such undertakings arguable. Profits are further eroded because thinnings are unsaleable in most countries, as secondary bush provides low-cost alternatives for the supply of poles and fuelwood.

It must also be remembered that plantations are high-risk investments, not only because of the often-cited pest and disease problems, but because discontinuity of maintenance caused by a lack of funds is the rule rather than the occasional exception. Thus, valuable and expensive plantations in Gabon, Liberia and Cameroon have largely gone to waste as a consequence of cuts in the budgets of their forestry departments.

Managed natural forest, on the other hand, does not require such continuous or high investment. Indeed in a recent study carried out by the Société de Développement des Plantations Forestières and the Centre Technique Forestier Tropical in Côte d'Ivoire, it was found that, over a thirty-year period, 1 cu. metre of timber derived from managed natural forest only needed an investment of US\$5.60, whereas 1 cu. metre of timber grown in a plantation required US\$7.40. The economics of management of the natural forest would improve further if a wider range of natural-forest species could be introduced to local markets. However, it must be remembered that investments in the natural forest are usually compounded for longer periods.[3]

In short, plantations are appropriate under some conditions and can be an important complement to natural-forest management, but they are a high-cost alternative and an incomplete substitute for the other values of natural forest.

Natural-forest management can be practised at various levels of intensity (see Chapter 1). At a minimum, it requires demarcation and protection, inventory, and the regulation and control of exploitation of the forest. More intensive management involves silvicultural interventions such as releasing and the regeneration and cutting of climbers. It might be necessary in countries such as Ghana and Côte d'Ivoire which have little forest left but want to sustain the present cut. The high costs involved can only be justified in the case of forests with a high economic potential – forests that are rich in exploitable timber and where the cost of transport to markets is low. In the African countries studied, few forests fulfil these requirements. This chapter mainly deals, therefore, with the conditions necessary for extensive management to be successful, and silvicultural interventions are only dwelt upon briefly.

It is argued below that there is hardly any genuine, sustained-yield forest management in the countries studied, whether extensively or intensively managed. Forest is destroyed at an ever-increasing rate by unsustainable farming practices, and the extraction of timber is accompanied by little or no reinvestment of benefits obtained from it in order to ensure the capacity of the forest for future production.

The reasons for this are found in the broader context in which the management of the forest is conducted: land-use and forest policy and legislation; the application of these; the economic circumstances conditioning timber exploitation; and the availability of the information needed to manage forests sustainably. These are discussed extensively later in this chapter.

PRESENT STATE OF NATURAL-FOREST MANAGEMENT AND EXPLOITATION

How much forest is being managed at present?

One can safely say that, at the present time in the six countries studied, there are no sustained-yield forest management systems which are being practised over large areas. But an extensive management system has recently been introduced on the SOFIBEL concession covering 80,000 ha in Deng-Deng Forest, Cameroon.

Natural-forest management was practised in Nigeria (Tropical Shelterwood System (TSS), 200,000 ha), Ghana (TSS and various selection systems on several hundred thousand hectares), Gabon (stand improvement, 130,000 ha) and Côte d'Ivoire ("Amélioration des Peuplements Naturels", 50,000 ha). Such systems have been progressively abandoned, however, perhaps with the partial exception of Ghana where at least some of the prescribed management activities are still being carried out in some of the forest reserves.

Recently, various measures necessary for good forest management have been taken in Africa's TMF zone. For example, in Côte d'Ivoire there is firmer forest policy; the Congo has adopted forest legislation conducive to sustained-yield forest management; Cameroon, a more efficient logging control system; and Liberia, higher forest fees that are reinvested in the forestry sector.

The future forest exploitation strategy chosen by most countries is one of extensive management of large concessions by highly capitalized logging and processing enterprises. Proposals to this end have been made in Côte d'Ivoire for half of the remaining productive high forest (700,000 ha); in Cameroon, where three 200,000 ha industrial complexes are foreseen in the south and south-east; and in the Congo ("Unité Forestière d'Aménagement 6", 200,000 ha).

Intensive management involving silvicultural operations is being attempted in various pilot projects. Examples of these are the 5,500 ha treated under TSS in Bobiri Forest Reserve, Ghana; the 10,000 ha that were subjected to release operations in Yapo Forest, Côte d'Ivoire; and the 5,000 ha (to be extended to 30,000 ha) treated similarly in the Forest Reserve of So'o Lala, Cameroon.

It should be stressed here that the experience gained in pilot projects, however valuable from a silvicultural point of view, is of only limited value for the establishment of operational management systems over large areas. Scaling up pilot projects to operational level involves not only quantitative but also qualitative factors.

What is happening to the tropical moist forest?

If no sustained-yield forest management is being practised over large areas in the African TMF, what is happening to the forest?

Originally, some 105 million ha of the six countries studied were covered with TMF; of this only 68 million ha (65 per cent) are now left (see Table 3.1). Deforestation has been especially heavy in the three West African countries, amounting to 50, 80 and 85 per cent of the original TMF area for Liberia, Ghana and Côte d'Ivoire respectively, whereas forest loss in Cameroon, the Congo and Gabon has been moderate at 22, 15 and 4 per cent. The figures should be treated as indications only, because deforestation data are highly questionable and include under one heading such different phenomena as conversion to permanent agriculture, and shifting cultivation with long and short fallow.

The causes of deforestation are diverse, but forest clearing for shifting cultivation is generally accepted to be the most important one. But, more often than not, it is a chain of events, started by the opening of a forest area for logging, mining or as part of general infrastructure development, and completed through land clearing by farmers, that leads to forest degradation and finally outright deforestation.

Extensive areas of forest have also disappeared through large-scale agricultural development (cocoa, coffee, oil palm, rubber).

Countries fall clearly into two groups: one in which most forests

Table 3.1: TMF resources in the six countries studied, 1988

Country	TMF act. (orig.) (million ha)		PHF area (million ha)	Annual loss (ha)	(%)
Cameroon	17.9	(23.0)	16.0	150,000	(0.9)
Congo	21.3	(25.0)	12.7	110,000	(0.9)
Côte d'Ivoire	2.0	(14.5)	1.5	no data available	(>2)
Gabon	20.0	(20.9)	18.0	15,000	(0.08)
Ghana	1.6	(8.1)	1.1	no data available	(>2)
Liberia	4.6	(9.6)	2.6	50,000	(1.9)

Note
Actual tropical moist forest (TMF act.) and productive high forest (PHF) area are in millions of ha, with, in brackets, the approximate original extent of TMF; annual loss is given both in absolute (abs.) terms (in hectares) and as a percentage of PHF. Productive high forest equals tropical moist forest minus areas that cannot be logged because of site conditions (permanently inundated, steep areas) or for legal reasons (national parks and other protected areas). Deforestation in Côte d'Ivoire and Ghana has recently slowed down in absolute terms because so little of it is left; percentage loss is still high.
Sources: Country missions and various.

have already been exploited, over-exploited or converted to agriculture (Ghana and Côte d'Ivoire in West Africa); and one where there are still large areas of relatively undisturbed forest left (Cameroon, the Congo and Gabon in West Central Africa), because population densities are low, access to the forest is difficult and ports are relatively far from the forest. Liberia seems to occupy a middle ground. However, these country-wide figures mask important regional disparities within countries; these are discussed below.

One of the consequences of this rapid forest loss is that, if present trends continue, Côte d'Ivoire and Ghana are bound to become net importers of timber in the late 1990s, following the example set by Nigeria in the 1970s. "Who will be next?" one is inclined to ask.

In the case of Ghana, forests will have to be managed more intensively if the annual cut is to be sustained, as there is little or no forest left to reserve. In Côte d'Ivoire, the annual cut has already declined rapidly over the last five years (from 4 million to less than 1 million cu. metres). In Liberia, it will not be possible to sustain the present annual cut unless more forest is reserved and protected against encroachment.

In the other countries studied, problems in sustaining the cut are not apparent from aggregate data for their forest resources. In the Congo, however, and to a lesser extent in Cameroon and Gabon, important infrastructural works will have to be undertaken to sustain the cut, as the areas at present providing the bulk of timber production are becoming progressively depleted.

As the diameters of available logs decrease and their quality declines, processing problems are expected to arise.[4] In Ghana and Côte d'Ivoire, grades have already fallen dramatically.

Much more important, however, are the serious fears which begin to arise about the environmental consequences of deforestation and forest degradation, not only in general terms, but also in their specific impact upon the viability of other land uses. Fears exist in Ghana and Côte d'Ivoire, and also in Liberia, about the Sahel spreading south and the harmattan winds penetrating the countries further and further, and about a likely decrease in those levels of humidity in the dry season which are essential for the cultivation of crops such as cocoa.

Thus, in summary, the forest is being destroyed by unsustainable farming practices, made possible by roads intended for logging, mineral extraction or general infrastructural development which provide ready access to farmers.

Timber exploitation and processing

What is the effect of logging on the forest itself? Two factors seem to have been decisive for the character of timber exploitation in the TMF zone in Africa: the distance of the forest from a suitable seaport for export; and the strength of local markets that can absorb the less valuable products.[5] Thus the forests of Côte d'Ivoire and Ghana, where ports are near and the domestic markets are stronger because of the large population, have been exploited much more intensively than other forests in the region.

This pattern of very selective exploitation in remote forest areas combined with fuller forest utilization in more accessible regions is also clear within national boundaries. Examples are the Congo (north east versus Mayombe and Chaillu), Zaïre (Cuvette versus Bas-Zaïre), Cameroon (Congolese forest in the south east versus the Biafran and semi-deciduous forests) and Gabon (centre and east versus the coastal zone).

Even in countries where little forest is left, differences are marked, as is, for instance, shown by the fact that *Ceiba pentandra*, a low-quality veneer timber, is being exploited in the coastal zone but not elsewhere in Côte d'Ivoire. This phenomenon illustrates the overriding importance of transport costs: it is not merely timber quality, but rather timber quality relative to harvest and transportation costs that determines whether or not a certain species will be exploited.

Despite these important regional differences, some general observations can be made. In Africa TMF exploitation has generally been very selective, with volumes of 5 to 35 cu. metres per hectare (and sometimes as little as 2 cu. metres) being removed, whereas in South East Asia, volumes exploited are 50 to 120 cu. metres and sometimes even more. This is not only due to the fact that African TMF is less uniform and less rich in commercial timbers, but also to those factors mentioned above, such as more difficult access and evacuation, and little-developed domestic markets. In the early 1970s, for instance, available rents per hectare of forest in East Kalimantan were fifteen times as high as those in Ghana.[6]

In general harvesting has been limited to few species and low volumes, but this situation is now changing, especially in the more accessible areas. Such factors as the disappearance of prime species by overcutting, bans on the export of the species with the best log form (in force in Ghana and Côte d'Ivoire and foreseen in the near future for Liberia), often combined with tax incentives, have greatly increased the interest of loggers in the so-called lesser-known species (LKS), some of which are now fetching quite good prices on the log market. A case in point is a species of *Aningeria*, little known before 1973, whose price is now on the level of prime species like sapele. Even so, the transport cost of these LKS, both to domestic and international markets, is often prohibitive, except in the most accessible areas.

This lightness and selectivity of harvesting have some important side effects. It is thought, for example, that the regeneration of commercial species and, in particular, the release of their advance growth are not encouraged by low-intensity exploitation, because these species are mostly "gap opportunists" which need much light to grow into the canopy. Also, in relation to the volume harvested, such selective exploitation opens up to shifting cultivators comparatively large areas of forest which were previously inaccessible to them. This situation, in combination with the general lack of protection of logged-over forests, has led to substantial deforestation in Cameroon, Côte d'Ivoire, Ghana, Liberia and the Mayombe region of the Congo.

Probably because logging has such a small effect, conflicts with local people are limited. Localized problems do occur, for instance in the east of Cameroon where concessionaires cannot log moabi, *Bailonella toxisperma*, for fear of a violent reaction from the local population who use the seeds of this species as cooking oil. Another case in point is the refusal of loggers involved in salvage fellings in cocoa areas in Ghana to pay proper compensation to the farmers affected. Little is known about the impact of logging on forest-dwellers such as the pygmies and other tribes living in Cameroon, Gabon, the Congo, the Central African Republic and Zaïre; but the widespread animosity to loggers, reputed to occur in South East Asia, of some forest-dwellers who see their livelihoods threatened by intensive logging operations, does not seem to exist in the African TMF zone.

On the contrary, interactions are often quite positive, with loggers providing employment, roads to markets, schools and dispensaries. The problem is rather that in the absence of any forest management the logger is bound to move on some day, leaving all these benefits to decay.

Most investments in the processing industry do not seem to have started with very serious intentions; they were rather accepted as a cost by timber exploiters in order to be able to continue their extremely profitable log exports.[7] This situation now seems to be improving greatly in Ghana, Côte d'Ivoire and, to a certain extent, in Cameroon and Liberia. Important constraints to industrial development are posed by the little-developed domestic markets in the Congo, Gabon, Liberia and parts of Cameroon.

FOREST EXPLOITATION AND MANAGEMENT: THE WIDER CONTEXT

Forest and land-use policy

In the African TMF zone, most of the forested land is in the public domain or at least nominally under government control. Good forest management, therefore, depends on the effective implementation of appropriate government policies.

As we have seen, people are destroying the forest resource at an alarming rate in many of the countries studied. This has often been ascribed to the high growth of population in Africa south of the Sahara, but at least part of the destruction is due to inadequate, and sometimes even counter-productive, government policies.

In drafting their forest legislation, most countries have committed themselves more or less strongly to policies of sustained yield. In practice, however, such policies are not pursued. One of the most important reasons is that, in the design of policies, forests are considered as a convertible rather than a renewable resource. In the political arena, agriculture is often perceived as more important than forestry and, even worse, as quite isolated from forests and forestry. Sometimes, forests are even seen as an adversary to development: "development starts where the forest stops".

Thus, even if forests are regarded as a renewable resource in forest legislation, the law is simply not applied: the forest protection envisaged in these laws is not enforced, financial resources for forest protection are not allocated, and offenders against the law are mostly acquitted – or if they are taken to court, the negligibly low penalties imposed do not really function as a deterrent.

Neither have foresters themselves been very instrumental in improving policies. In former times, their common preoccupation was with three vital factors: the needs of the population and of posterity; the allocation of land for agriculture; and the indirect benefits of forestry, particularly its role in protecting the environment for humankind. As control over the resource used not to be a major problem, they operated in a relatively isolated way, concentrating on forest reservation, on the biological characteristics of the forest and on the demands of mostly foreign markets.

But, recently, the environment of forest management has become much more complicated. Conflicting claims on forest lands have multiplied, and forest management has had to become much more sensitive to the needs of local populations and to the local and national political environment. These changes have found the foresters ill-prepared. Furthermore, rapid changes in available technology, not only in harvesting and processing but also in means of communication, continually change the relevant circumstances.[8]

There has been no acknowledgement of the rate of change in the environment for management, and little effort has been made, either by foresters or planners, to integrate forest policy with the policies of other sectors such as agriculture and urban development. Statements of policy have not faced up realistically to the interactions between the management of the forest and such factors as population increase, urbanization and production from a limited land resource; while, in the meantime, urbanization combined with population growth has caused an enormous rise in demand for foodstuffs, which have had to be produced extensively, consuming large amounts of forest land.[9]

Assuming that the process of opening up the forest as described above is going to continue unabated, then the future of the TMF resource depends on changes in agricultural and general economic development policies.

These will determine whether or not people will continue to be "pushed" into the forest, and whether shifting cultivators can be helped to pursue modes of land use that consume less high forest, thus reducing the "pull" into the forest. In other parts of the world this is very explicitly a political problem, as is demonstrated by the unequal distribution of fertile agricultural lands in most Latin American countries.

In Africa, political landlessness is less of an issue, and intensification of land use is hindered not only by a lack of well-designed government policies but also by extremely limited availability of capital, little-developed domestic markets for raw materials, poor infrastructural development, and low labour availability in rural areas due to pronounced rural–urban migration especially among young people. The resulting process of land extensification implies the wasteful use of remaining forest resources.

The need for integrated land-use planning is being recognized more and more. In the management plan for the Unité Forestière d'Aménagement 6 in the Chaillu region in the Congo, and for the So'o Lala Forest Reserve in the south of Cameroon, forests along the roads have been set aside for future agricultural expansion. Various forms of multiple-use management, combining conservation and rational utilization, are being sought in an IUCN programme funded by the European Development Fund.

Forest legislation and reservation

In all the countries studied, reservation of part of the TMF is foreseen. In Cameroon, Congo, Côte d'Ivoire, Gabon and Ghana the legislation provides for reservation of forest for various uses: permanent production forest, protection forest, different types of nature reserves, and sometimes even recreation forests. In Liberia, the 1953 Forest Act provides for the creation of forest reserves and national parks, but does not distinguish between production and protection forest.

Despite all these provisions, the actual situation looks rather bleak. Although dereservation *de jure* is rarely encountered, encroachment on forest reserves by farmers, and sometimes even their complete occupation, is the rule rather than the exception in Ghana, Côte d'Ivoire, Cameroon and Liberia. Since hardly any protection and production forest reserves have been established in Gabon and none at all in the Congo, and population densities are low, these countries do not face the same problems at present.

Forest reserves in the countries studied do not generally cover extensive areas, and their integrity has often been severely affected (see Table 3.2). Furthermore, not all reserved TMF is productive high forest, as figures cited include protection forests on steep slopes and critical watersheds.

In most countries, the legislation is vague about the definition of the various types of forest reserves, and no indications of areas or criteria other than "adequate" or "necessary" exist. Where reservation targets are more clearly stated (20 per cent in Cameroon, progressive reservation of all remaining forest in the Congo), they have not been reached.

There are a number of reasons for the slow progress in reservation programmes common to all or most of the countries studied, though the degrees of importance vary:

- the unpopularity of forest reservation with local populations, authorities and loggers
- the fact that there is little co-operation, or sometimes even outright hostility, between the agricultural and forestry sectors
- the complicated and lengthy procedures involved; local and provincial authorities, and various government departments (ministries of lands, agriculture, environment, mines, scientific research, tourism), at best not particularly interested in but more often opposed to forest reservation, have to be consulted
- the lack of staff and, even more important, of means and funds to carry out reservation programmes
- the too rapid circulation of staff in key positions in the forestry departments; reservation is a long-term effort, involving lengthy negotiations that suffer enormously from frequent personnel changes
- the doubtful reliability and dispersed character of inventories completed make it hard to defend reservation against other interests.

Table 3.2: Forest reservation in the countries studied, 1988

Country	Pres. (add.)		% under TMF	Annual cut
Cameroon	1.3	(1.3)	60	2.0 ('87)
Congo	0.0	(20.0)	–	0.7 ('86)
Côte d'Ivoire	3.0	(0.0)	40	0.9 ('86)
Gabon	0.2	(2.0)	100	1.3 ('87)
Ghana	1.7	(0.0)	75	0.8 ('86)
Liberia	0.7	(0.6)	90	1.0 ('87)

Note

Both present (pres.) and planned additional (add.) forest reserves are given in millions of ha; the second column gives an estimate of the percentage of forest reserves actually under TMF, and the third column lists the most recent figures available on annual log production in millions of cu. metres. Forest reserves include protection forest for which there are often no separate data available, but exclude national parks.

One of the main barriers seems to be a widespread popular resistance against forest reservation that has never been dealt with adequately. Already, in colonial times, the winning of local support for reservation was sometimes badly neglected (although Nigeria and Ghana seem to be exceptions to this), and the situation has hardly improved since. Much too little has been done to gain the confidence of the elders and the understanding of the young. As a result forest reserves were and are seen as a liability and a nuisance at the local level and, when the countries became independent, some gazetted forest lands were immediately handed over to farmers. Often, forest reserves are still seen as a remnant of colonial rule.[10]

In most countries, provisions for sustained-yield management are made in the legislation. In the case of the Congo these are quite detailed, and the possibility thus created for specific regulations to be applied to each "forest management unit" is a prerequisite for sustained-yield management. In other countries, such provisions are made in more general terms, but sometimes the decrees specifying the precise conditions to be respected are being held up at various levels, as in Gabon and Liberia.

It can generally be said, however, that problems concerning forest legislation are due not to its inherent inadequacy, but to the fact that it is not consistently applied. General rules for forest resource management such as reservation of adequate production and protection forests, and sustained-yield management of the forest resource as a whole, are present in the legislation, but they are not enforced. Rigorous application of forest legislation depends on firm government policy, as discussed previously.

Nevertheless, there is certainly some room for improvement in existing legislation. As we have seen, prescriptions are often made in rather vague terms, and would benefit greatly from rigorous definition and specification. More fundamentally, forest legislation needs to recognize the links between the various land uses and users, and it will have to integrate and define clearly such new notions as integrated rural development, multiple-use management, community and village forests, etc.[11]

The control of logging

Management implies controlling logging and, in some instances, imposing certain restrictions on it. Indeed, the quality of timber harvesting operations is all-important for the future potential of the forest, especially in TMF in Africa, where forest heterogeneity and low commercial timber volume preclude the use of mono-cyclical systems. Consequently, the controls and restrictions have to be realistic, not inflicting too much financial loss on the logger; for, if they do, he will not respect them. Retention of seed-bearers (leaving one tree per hectare out of every two or three exploitable hectares) is an example of a measure which greatly affects financial viability and may not even be necessary.[12]

Most logging is subject to concession agreements of one kind or another. In these, the duration of the agreement, the boundaries of the area to be logged and the level of taxation are fixed, and rules of conduct for loggers are given. Those rules generally include several or all of the following elements:

- standards for the quality of logging roads (maximum slopes, width, drainage) and their maintenance
- the obligation to carry out a 100 per cent pre-exploitation survey and mark trees before felling
- the obligation to respect minimum exploitable diameters and maximum slopes, and to limit felling damage
- the obligation to protect coupes after felling has been completed.

Furthermore, demands for other infrastructure (schools, dispensaries) are sometimes listed.

Detailed instructions about measures to limit felling damage, such as climber cutting, directional felling and marking of residuals, are not generally encountered, with Ghana being among the notable exceptions. Also missing is the obligation to carry out an audit of regeneration before felling and a post-exploitation inventory followed by continuous monitoring of the developing stand.

It has to be concluded therefore that quite a few of the elements that need to be part of the logging process, if forest management is to be sustainable, are not even mentioned in the present concession agreements. Furthermore, the absence of control or the unlikeliness of its occurring does not induce the loggers to comply with existing rules and regulations.

A factor further complicating effective control of logging operations is the widespread practice of third-party logging, whereby concessionnaires, often nationals without any logging equipment who have obtained their concession as a political favour, let the permit to other operators in exchange for a fee.

Conditions for the granting of operating licences (such as the concessionnaire having paid all due taxes and complied with all other pertinent regulations) are often waived. In Ghana for instance, only 31 per cent of all concessionnaires were found to be operating with a valid property mark in 1986. The situation seems to be similar in some of the other countries studied.

The basic instruments that are supposed to ensure the future productivity of logged-over forests are the minimum exploitable diameters established for the various species and the length of the felling cycle for the forest as a whole. These should therefore be based on factors such as site productivity, and the increment rates and regeneration requirements of the species involved. In practice, however, they seem to be based on short-term financial considerations rather than on any biological or sound economic arguments. The length of the cutting cycle, for instance, varies from twenty-five years in Liberia to fifteen years in Ghana, with most foresters interviewed conceding that the real biological cycle is probably nearer forty years! In countries such as Côte d'Ivoire, there has been no cutting cycle at all, and exploitation has been a matter of sheer anarchy.

In some cases, the legally fixed cutting cycle does not equal the actual one, because low area fees have induced concessionnaires to include farm and other non-forest lands in their concessions. In Liberia, the 4 per cent annual coupe allocation has been based on total concession area rather than the productive forest area in it. This has resulted in a *de facto* reduction of the felling cycle under some forest products utilization contracts. If for instance 50 per cent of a concession is inaccessible or is being farmed, the actual felling cycle is reduced from 25 to 12.5 years. As a matter of course, such practices render the planning of sustained-yield timber harvesting impossible, and increase the incentive for excessive logging in the exploitable forest.

The story is much the same for minimum exploitable diameter limits, varying from 1 metre for prime species in Liberia and Ghana, to 1.0 or 0.8 metres in Cameroon and 0.6 metres in Côte d'Ivoire for the same species. Okoumé, which like many veneer species is known to double in value in the seven years after it has reached 0.7 metres diameter, has nevertheless a minimum exploitable diameter of 0.7 metres in the Congo and Gabon, and 0.6 metres in Equatorial Guinea.

Concession size and duration equally pose problems. The minimum size at which a concession can be managed effectively and sustainably is estimated by foresters in Ghana to be about 5,200 ha (40 years (felling cycle) times 130 ha (compartment size)), but areas below 800 ha are given out in concession. The duration is rarely longer than one cutting cycle. Concessionnaires cannot be expected, therefore, to show an interest in the future productivity of the forest resource.

On the other hand it has equally proved wrong to suppose that sustained-yield management will be immediately brought about by providing concessions of adequate size and duration. This is demonstrated by widespread unsustainable logging in Africa prior to independence, when concessions were large and their duration fifty years or more.[13] Many other conditions, which are often of a socioeconomic nature, also need to be fulfilled.

In some of the countries studied, such as Ghana, the Congo and Gabon, it was originally intended that the forestry department would do the 100 per cent inventory prior to exploitation. However, the means to carry out this task are not available, and it has therefore often (except for some forest reserves in Ghana) been delegated to the loggers, which is not unreasonable. But, in the present situation where any control of the inventories done by the loggers is impossible, this clearly opens the possibility of extreme high-grading. Comparisons between inventories done by loggers and by the Forestry Department in the Congo indicated that the latter obtained commercial volumes up to three times higher than the figures handed in by the loggers.

As markets for all but the fifteen or so well-established African timber species show extreme fluctuations, the forest authority usually grants some flexibility to loggers in working their coupes. Re-entry into previously logged coupes before expiration of the regeneration cycle is sometimes permitted, when a demand develops for a previously "unmarketable" species that has been left standing. However, this may endanger regeneration, reactivate climber activity and increase damage to advance growth and soil. It also involves substantial dissipation of potential timber rent due to the inefficiency inherent in repeated passages of heavy machinery through the forest.

In Ghana, repeated re-entry as and when loggers obtain sales contracts for lesser-known species is now almost standard practice. In Côte d'Ivoire, there is no such thing as a regeneration cycle anyway, and forests have often been logged over three times in ten to fifteen years. Until very recently, re-entry was also common in the coastal okoumé forest in Gabon, where it is known to be tremendously wasteful. Stands where forty mature okoumé trees per hectare could be harvested in fifteen to twenty years were being completely destroyed to take out the two to three trees that had reached maturity already. In Liberia, provisions for re-entry exist only in the case of previously truly unmarketable species, and has therefore rarely been requested.

In Cameroon, some flexibility is granted to the loggers by allowing them to keep coupes open for a maximum of three years, after which they are in principle closed for a full felling cycle. This is probably less harmful than re-entry as practised in some of the other countries.

Salvage permits allow loggers to remove all merchantable timber, regardless of any minimum exploitable diameter limits or other restrictions. These permits are in principle only given where conversion to agriculture or other non-forest land use is imminent. Salvage logging in general involves the removal of large quantities of timber per hectare and leaves the forest severely degraded, often unable to regenerate itself. On the one hand, salvage permits are a useful tool to prevent wastage of timber resources in forest lands to be cleared for agricultural or other purposes. Estimates are that for instance in Côte d'Ivoire 200 million cu. metres of merchantable timber have simply been burnt during unplanned land-clearing operations over the last twenty-five years. Equally, the Ghanaian Forestry Department estimates that only 15 per cent of forests were properly exploited before they were converted to other land uses.

However, salvage logging is not an undivided blessing for two reasons. First, logging companies wielding political power may obtain salvage permits for areas that are not planned for conversion, sometimes even within forest reserves, as was observed by this author in N'krabia Forest Reserve in Ghana. Also, salvage logging is permitted half a mile on either side of logging roads in Liberia, and is used as an incentive for loggers to undertake various infrastructural works in Gabon.

Second, the fact that timber below minimum diameter becomes available for export or processing complicates the control of timber exploitation tremendously, although this problem seems to have been solved in Liberia by separating salvage and concession logging in time and carefully controlling property marks. But, especially where funds available for control missions are scarce (the Congo, Gabon, Côte d'Ivoire, Ghana), the possibilities for abuse are enormous.

In recent years problems have arisen due to anarchic felling of trees with power chainsaws: this is reported to be increasing in Liberia, Ghana and Cameroon. In Cameroon the present policy is to tolerate this practice, and try to combat the worst excesses rather than crack down on it altogether, which would be unrealistic given present control capacity.

An interesting experiment with a new logging control system has been started in Cameroon. This country has suffered much in the past from logging of trees below minimum exploitable diameter, disrespect of concession boundaries, the abandonment in the forest of parts of logs and "cut but suspended" trees, and other violations of logging licence conditions. In this it is no different from many of the other countries studied. As a result, a centralized and computerized system for controlling logging was tried in 1988 in the Southern province and will be extended to cover the whole moist forest zone in 1989.

A number of important innovations have been introduced. These are:

• all relevant data are fed into one computerized information system: recognized logging companies and their property marks, surface area and situation of concessions, volumes harvested, taxes due, (non-)compliance with existing regulations on logging and processing, etc.

- all volume and tax calculations are carried out by the central computer, liberating forestry staff to control logging offences on the ground
- in a bid to prevent corruption, forest guards will no longer be attached permanently to logging companies, but will participate in regular but unannounced control missions instead
- the Forestry Department has independent teams to control the quality of the regional forest guards' work.

Preliminary results of the new control system seem to be quite encouraging.

Socioeconomic and financial policy

Underlying traditional government policies that favour development of the tropical forest is the basic belief, implicit or explicit, that forest industries can be the mainspring for wider economic development, beginning in the underdeveloped rural areas but ramifying throughout the national economy.[14]

General studies of industrial development suggest, however, that the assumed connection between development of the forest sector and national economic development does not operate in actual experience.[15] However, in the absence of specific, systematic research in this area, the development hypothesis remains at the centre of forest policy in many developing countries, whether justified or not.

Confronted with tremendously selective and wasteful timber exploitation, most of the countries studied have tried to improve the economics of forest utilization, both by extending the number of species harvested, and by promoting local timber processing.

The utilization of lesser-known species

The virtual stagnation of domestic timber markets in sparsely populated countries such as Gabon, the Congo and Liberia, but also in countries struck by economic recession as Cameroon and Côte d'Ivoire are at present, has continued to tie forest industries too closely to the export market with its emphasis on high quality and extreme selectivity. While logs of the so-called lesser-known species

(LKS) can be promoted for export, albeit with great difficulty, the lack of a domestic market for unexportable processed timber of lower and medium quality is a serious inhibitor of forest industry development. Where these circumstances coincide with the existence of substantial areas of unexploited forest, as in the Congo, Gabon and to a lesser extent Cameroon, the dominance of a highly selective export market is likely to continue for several decades.

In all countries studied, incentives have been created to promote the utilization of LKS.[16] In some countries, such policies are quite successful, although increased utilization of LKS can be just as much a sign of resource depletion (the established timbers having been overcut) as an indication of effective policies.

Still, in most countries one to three well-established species provide 50 to 80 per cent of the total log volume harvested, and exploitation remains selective, especially in remote areas. Incentives are mainly in the form of differential taxes and of legal obligations that prescribe a certain minimum percentage of LKS as part of the total cut. But the differences between the taxes and their absolute value are often not big enough to have any influence; moreover, the minimum percentages are often not respected, on account of negligibly low fines or the fact that appropriate fines may be imposed but are not recovered, as is the case in the Congo.

Côte d'Ivoire, assisted by the Centre Technique Forestier Tropical (CTFT), has undertaken a most spectacular campaign to promote the export of LKS: in 1979 1.47 million cu. metres, or 53 per cent of timber exported, consisted of twenty-five LKS that had not been exported before 1973. This effort has also benefited neighbouring countries, although it has done little to solve the problem of increasing domestic utilization.

Utilization of lesser-known species is often considered to contribute substantially to better forest resource use and conservation. This seems to be a contentious matter, and problems do occur, such as the re-entry into logged coupes, described above, in order to harvest LKS. At the very best, it can be stated that whether the use of LKS will have a positive influence is conditional upon a number of other factors, such as the strict application of forest legislation and concession clauses.

In-country processing
Often, as a token concession to authorities, timber has been processed with derelict equipment causing tremendous waste, while in

reality log exports were still all-important. At present, this situation seems to be improving as more and more enterprises start to take processing seriously, especially in countries where the forest resource is getting thin and demand for timber concessions is greater than supply.

If serious processing is indeed the aim of investment, there are other problems. Manufacturing processes tend to be capital-intensive and the costs of production are sensitive to economies of scale. Markets for these products are international, and success in either import substitution or export is subject to many external influences. The industries require high inputs of energy, management skills, spare parts and some raw materials such as chemicals that may not be available locally. It is not without risk to encourage timber-processing industries, and the ratio between jobs provided and the amount of capital invested is often extremely low when compared to those common in other sectors.

It has been stated that governments often encourage inefficient processing while themselves incurring losses in terms of revenue forgone from log exports and from "free rides" such as export-tax exemptions and income-tax holidays.[17] A case in point is Cameroon, where 3 cu. metres of logs are used to produce 1 cu. metre of processed timber which has a value equivalent to that of 2 cu. metres of logs! This precarious situation has no doubt been aggravated by the high tariffs imposed on imported processed timber, especially plywood, by Japan, Korea, Taiwan and, to a lesser extent, the EEC and USA, despite some positive clauses in trade agreements such as GATT and Lomé.

Forest fees and taxes
Fair forest fees and taxes can create important incentives for loggers to manage their concessions sustainably; they can, at the same time, provide government forestry departments with the means to control and manage forest exploitation. If taxes are too low, loggers will often revert to rent-seeking behaviour, trying to reap the biggest possible profit in the shortest possible time with minimum investment. This can result in excessive harvesting either through more rapid exploitation of existing concession stands or more extensive harvesting of virgin stands, or both. On the other hand, if taxes are too high, the profitability of logging enterprises is endangered and disruption of economic activity in the forestry sector can result. The latter phenomenon, however, does not occur in any of the countries studied.

The taxation of forest products varies tremendously among the countries studied. Important characteristics are whether taxes are directly related to FOB prices, whether they are fixed in local currency and therefore inflation-prone, whether differences in taxes provide incentives for utilization of LKS and increased processing, and whether or not high-grading is effectively discouraged. The efficiency of tax collection is another central issue: there are large differences in the rate of recovery of taxes, and in the degree to which forestry authorities have direct access to the fees levied.

High area fees discourage high-grading and completely prevent marginal forests – whose non-timber benefits are probably more important than their timber yields – from being logged at all. They are therefore conducive to sound forest management. But area fees are generally very low in the countries studied and are usually supplemented by taxes on timber harvested. In Cameroon forest fees are only charged on an area basis, which greatly complicates the control of product flow.

Fees on timber harvested are sometimes levied per tree felled as in Ghana, but mostly on volume felled. Fees charged per tree would encourage high-grading more than would volume-based fees, because less-than-perfect specimens of commercial species are left standing if the logger has to pay the same fee for felling them as for perfect specimens. This only holds of course if fees are high enough; fortunately they are so low in Ghana that they do not have this effect.

Volume fees tend to be little differentiated, a ratio of between 1:3 and 1:6 being common between preferred and non-preferred species. Reforestation fees are often imposed as flat rates per cubic metre harvested, independent of timber value. In principle these practices should encourage high-grading, but they are often not high enough to do so in practice. Also, for most non-preferred species, it is extraction costs rather than forest fees that are prohibitive.

Taxes on processed products or logs to be used for in-country processing can be imposed in two ways: on logs before they enter the plant – this creates an incentive for more efficient processing but also for wastage and the abandoning of logs in the forest, as well as for smuggling and illegal log exporting; and on processed products leaving the plant, the more common practice, which does not provide any incentive for more efficient processing. The former mode of taxation is dependent on a well-functioning logging control system on the ground, which is exceptional in the countries studied.

Liberia could function as a model for the other countries studied, as it seems to have one of the most active taxation policies, which differs from those elsewhere on a number of points:

- the differential in log export tax, here called industrialization incentive fee, is about 1 to 40 (from US$58.56 per cu. metre for sipo to US$1.44 per cu. metre for low-value species), thus offering an important incentive for the utilization of LKS. Loggers removed as much as 65 per cent of exploitable volume in the 1986-7 annual coupe, which is high considering the quality of roads and infrastructure
- the concession rent, called land rental fee, has been increased recently from 10 to 25 cents per acre and per annum, or US$0.63 per hectare per year. This is much higher than in the other countries, giving some further incentive for fuller forest utilization but, more important, inducing loggers to give up the farmed and inaccessible parts of their concessions, which they were holding to increase their annual coupe of 4 per cent
- the stumpage fee consists of a cess-tax of US$0.40 per cu. metre, and a conservation tax (of US$3 or US$1.5 per cu. metre for class A and class B species), funds which are designated for forestry training and forest protection respectively and which are collected and used directly by the Forestry Development Authority. The conservation tax had helped to reopen 3,000 miles of National Forest boundaries by 1987.

Allocation of resources to government forestry institutions
Governments look at their tropical forest resource and its administration in a number of different ways. The forestry sector may be perceived as a source of revenue and foreign currency, as a source of employment, as a self-financing part of the economy or as a combination of these elements. The governments of all countries studied treat the forestry sector mainly as a source of revenue and foreign currency, although the provision of employment plays an important additional role in Ghana, and the Forestry Development Authority (FDA) in Liberia has been enabled to generate part of its budget through the independent levying of taxes.

Obviously, there is an enormous difference between government forestry institutions, able to generate their own budgets, and institutions that depend on the whims of ministries of finance and the like. Equally obviously, little output can be expected from forestry staff who have not been paid any salary for months on end, a common situation in the latter case.

Lack of staff does not seem to be the major problem, but rather the fact that those staff in place lack the means to carry out the tasks assigned to them. In Gabon, for instance, the Forestry Department's budget for current expenses was reduced from 700 million CFA francs in 1984 to 177 million CFA francs in 1988. This has resulted in few control missions and general inactivity.

Research on forest management is one of the areas that often suffers from a lack of funding. As forests are only seen as a source of revenue, the political priority attached to research and other investment in future productivity is extremely low. Forestry research is virtually non-existent in Liberia and Gabon, although some research is being done in foreign-funded projects; the infrastructure still exists in Ghana, but most senior researchers have left. In Cameroon there is some forestry research but it is dispersed over a number of institutions, mostly underfunded, and there is often little communication between researchers and forest managers about the results obtained.

Choice of loggers

Good mechanized forest exploitation requires large capital inputs to obtain the necessary machinery. This mostly puts national entrepreneurs at a disadvantage compared with foreign interests. In some countries, whether or not under political pressure to make logging a more accessible business to nationals, the forestry authorities are waiving forestry legislation and regulations in the case of certain national entrepreneurs, such as the "piétistes" in the Congo and the "coupes familiales" in Gabon. These loggers, however, do not have the equipment necessary for rational forest exploitation and often revert to destructive practices such as creaming off the forest along roads and cutting all seed-bearers even below exploitable diameter.

In other countries such as Liberia, it is simply accepted that foreign interests are bound to dominate the logging scene for years to come, rather than a few wealthy citizens wasting the country's forest resources, as an employee of the Forestry Development Authority put it. In still others such as Cameroon, action is being undertaken to support nationals who seriously want to engage in logging. This approach seems far more conducive to sound forest resource management.

Loggers are rarely silviculturists and, even if they were, they would have little incentive to behave as responsible technicians in the present situation. As one official stated: "If loggers see a merchantable sapele at 50 yards, they'll take the bulldozer to it in a straight line, no matter if that means destroying a young stand of the same species, because they do not recognize them anyway. Most of them are just timberjacks."

This means that important training activities will have to be undertaken if loggers are to be successfully involved in forest management. Training programmes are under way or foreseen in various countries, such as the PME/PMI (small and medium-sized enterprises and industries) programmes in Cameroon and Congo.

Economic circumstances conditioning forest management

In most developing countries, there is a dire need to generate foreign currency: exporting timber is one of the readiest ways to do so. As the prices of most other commodities (oil, gas, cocoa, coffee, iron ore and various other minerals) have steadily fallen in recent years, the pressure on the forestry sector to generate more cash has increased. This pressure is often expressed in completely unrealistic production targets for log production such as are found, for instance, in the five-year plans in the Congo (2 million cu. metres in 1990), Cameroon (3 million cu. metres in 1992) and Zaïre (6 million cu. metres in 2000). Such figures often do not take into account the limited productive capacity of presently accessible forest and the expensive infrastructural improvements (roads, rail, handling and storage facilities in ports) needed to bring about this kind of increase in production. Inherent in this sort of "planning" is a very real danger of overcutting easily accessible forests.

The figures in Table 3.3 are meant to give a picture of the pressure on the forestry sector not only to generate foreign currency, but also to relinquish forest fees and taxes in principle designated for forest resource management to the general state budget. Often, export revenues from timber are even more important than is indicated in this Table, as for instance in Liberia, where royalties for iron ore mining and rubber production have been paid years in advance for debt servicing, and in cocoa-producer countries where exporting cocoa is presently costing money due to depressed world markets.

In some countries non-exchangeable and often overvalued currencies make credit facilities very difficult indeed to obtain. Getting spare parts is problematic because there is little foreign currency to purchase them and because of a bias in shipping routes against ports where demand for cargo is irregular or availability of worthwhile return cargo not assured. In others, such as Liberia, the existence of a fixed and guaranteed exchange rate for a northern currency has caused enormous problems. There, timber exports to the European market (85 per cent of the country's exports) fell dramatically when soaring interest rates forced the US dollar ever higher in the early 1980s.

Table 3.3: Debt and timber exports in the six countries studied

Country	Foreign debt	Export value	Timber exp. abs. and %
Cameroon	2.8 (1986)	2.17 (1985)	92 (1987), 5–7% (1984-86)
Congo	1.9 (1985)	1.09 (1985)	40 (1986), 5–7% (1984-86)
Cote d'Ivoire	8.3 (1987)	2.88 (1985)	238 (1984), 9% (1984)
Gabon	0.7 (1985)	2.02 (1984)	143 (1986), 5–6% (1984-86)
Ghana	1.6 (1986)	0.63 (1985)	55 (1986), 8% (1986)
Liberia	0.8 (1984)	0.44 (1985)	65 (1987), 15% (1986-87)

Note
Foreign debt and export value are in billion US dollars, except the figures for Liberia, which are in billion Liberian dollars; timber export figures are given in absolute figures (in million US dollars) and as a percentage of total export value.
Sources: *Africa Review*, 1988, and country missions.

In addition, some very important economic factors are completely outside the control of the governments concerned. The general decline in the world's tropical timber markets in 1982-4 was due to a global economic recession.

Information needed for forest management

Good forest management is dependent on land-use and economic policy and forest planning. The quality of these depends on research: on policy, social and economic factors, botany, ecology and physiology.[18] But research is not a political priority in any of the countries studied. "Short-term concerns and interests prevail", as one of the participants at the seminar concluding the African part of the study said. Even so, government policy makers need more and better information; the reliability of the various figures quoted in this document, for instance, is variable, to say the least.

But, contrary to what is generally stated, the lack of available research results need not lead to inactivity. Management, if monitored, can itself be used as a tool of research. Ivorian experience has shown that research can be conducted while management is being practised and that results can be obtained in the relatively short time of four to ten years. Flexibility in the design of management projects, and monitoring of results, are of the utmost importance.

Economic policy research

As has been stated before, government economic policy regarding the TMF is often based on inaccurate information and is therefore flawed. Research is needed on the economic relationships between forest-sector development and overall economic development (not only involving timber, but also non-wood forest products, environmental and ecological benefits); and on appropriate, valid standards for choosing among policy alternatives.

Research on forest utilization potential

So far, static forest inventories concentrating on presently exploitable volume have ruled the day. These are useful but should be complemented by reliable data on forest dynamics, especially on increment and regeneration.

Extensive areas of TMF in the countries studied have been the subject of inventories (see Table 3.4); but the present usefulness of the results of those inventories is highly questionable, because they have been partly outdated by clearance for agriculture and timber harvesting and, to a lesser degree, by natural increment. Furthermore, many timber species not included in inventories have subsequently become acceptable for commercialisation. New inventories are, therefore, foreseen for most countries, especially the ones that have suffered the most rapid deforestation, such as Côte d'Ivoire and Ghana, where complete inventories of the remaining forests are under way or being planned.

The inventories that have preserved some of their usefulness are mainly those that were carried out in remote areas where clearing has been marginal, and logging, if any, was very selective.

Inventories are very expensive. Since most forest departments no longer have enough funds to put at the disposal of their inventory services, these activities often depend wholly on a foreign donor. Thus, inventories are being carried out in Cameroon with Canadian assistance, in the Congo helped by FAO/UNDP, in Côte d'Ivoire with French, in Ghana with British, and in Liberia with German assistance. The forest inventory service in Gabon is at present reduced to inactivity because it has no such assistance.

Table 3.4: TMF inventories in the six countries studied

Country	Area (period) (million ha)	% of PHF area	Ongoing/planned (million ha)
Cameroon	11.0 (1970s-80s)	66	4.0 (1990s)
Congo	3.5 (1960s-70s)	26	0.0
Côte d'Ivoire	15.0 (1960s)	0	1.5 (1990)
Gabon	6.1 (1960s-70s)	31	0.0
Ghana	8.0 (1950-65)	19	0.5 (1988)
			0.6 (1990)
Liberia	2.0 (1960s)	19	2.6 (1990s)

Note
The first column gives the area of completed inventories, the last column ongoing and planned additional inventories. The percentage of productive high forest area is an estimate based on those inventory results that still give an accurate image of present resource potential, expressed as a percentage of the total exploitable forest.
Sources: FAO, *Management of Tropical Moist Forest in Africa*, FAO Forestry paper (Rome: FAO, in press); country missions.

From a silvicultural, botanical and zoological point of view, the TMF resource is even less known. Forest inventories concentrate on stem volume of commercial timber species, and often contain little information on forest structure, small-sized trees, saplings and seedlings, and even less on non-timber biological resources.

Many forest inventories, too, are limited to providing static information about the resource; information on increment and regeneration is much scarcer, but is needed for good forest management. Many hundreds of permanent sample plots have been established to yield these kinds of data, but often they have been converted, or measurement has been discontinued, due to a lack of funds. In Ghana as well as in Côte d'Ivoire, many silviculturally treated and control plots are (still) being measured; such activities have recently been started or resumed in Liberia, Gabon, the Congo and Cameroon.

Silvicultural systems
Silviculturists in Nigeria and Côte d'Ivoire had experimented with natural regeneration and line planting for much of the first half of the twentieth century. Many other African forestry departments tried to take up the challenge of silviculture in moist

forest beginning in the 1950s. The main methods founded on natural regeneration will be briefly discussed here.

The *Tropical Shelterwood System* (TSS) was designed in Nigeria on the basis of tests which had been carried out for twenty years. The objective was to enhance the natural regeneration of valuable species prior to exploitation by gradually opening up the canopy (poisoning of undesirable trees, cutting of climbers) to obtain at least 100 one-metre-high seedlings per hectare over five years. The forest thus worked was logged in the sixth year; then cleaning and thinning operations were carried out over fifteen years.

The Nigerian Forestry Department treated 200,000 ha of forest in this way between 1944 and 1966, when the method was given up. The main problems encountered were the exuberant spreading of climbers following the opening up of canopy and the failure of the seedlings of valuable species to grow adequately. Besides, the poisoning of undesirable species eliminated trees which later turned out to be commercially valuable, e.g. *Pericopsis elata*.

Some Ghanaian foresters claim that TSS needs to be re-evaluated in the light of the increased marketability of new timber species, as too much emphasis was placed on red Meliaceae timbers in previous evaluations. Indeed, stands treated under TSS in Bobiri Forest Reserve in Ghana, show impressive stocking and growth rates of timbers such as *Piptadeniastrum africanum, Triplochiton scleroxylon* and *Terminalia ivorensis*.

In 1950, the Forestry Department of Côte d'Ivoire adopted APN. a technique related to TSS. It did so because initial results of the TSS in Nigeria seemed appealing, and because increasing domestic timber consumption demanded more geographic dispersal of activities and more species. The *Amélioration des Peuplements Naturels* (APN) method was applied from 1950 to 1960 on large areas of forests which had been logged over and were well stocked with valuable trees of average size. The aim was to favour the growth of these average stems and also to ensure regeneration through natural seeding of the valuable species by removing climbers and opening up the canopy. It was abandoned in 1960 when results were judged to be disappointing.

A *Selection System* has been applied in Ghana since 1960. Its objective is to ensure regeneration of forests well stocked with valuable species. Harvesting was meant to occur about every twenty-five years, after the Forestry Department marked the stand to retain well-distributed seed trees. This was to be followed by thinning operations. In 1970, however, the felling cycle was reduced temporarily to fifteen years in response to public allegations that over-mature timber was going to waste in Ghana's forest. The result has been considerable felling damage; regeneration has not been judged good, and less valuable shade-tolerant species dominate because of insufficient opening up of the canopy. Plans now exist to re-establish a longer harvesting cycle of around thirty years.

One management method that did not have as its direct objective the enhancement of natural regeneration was that called the *Improvement of Stand Dynamics*. This technique was used in Gabon in the *Aucoumea klaineana* forest to accelerate the growth of stems of valuable species of all sizes in naturally well-stocked stands, without particularly trying to provoke regeneration through natural seeding. The species grew in patches or clumps presumably due to natural seeding of forest trees in clearings or gaps. The objective was to let these stands attain commercial diameters as quickly as possible through thinning operations, but the production gain was never measured. After thus treating about 120,000 ha of forest (containing about 12,000 ha of pure okoumé stands), the Forestry Department gave up the technique in 1962 to switch to plantations of okoumé.

Recently, some promising initiatives in the field of forest management have been taken up. In 1976, an important trial was set up in Côte d'Ivoire by the Société Ivoiriennne de Développement des Plantations Forestières (SODEFOR) with the technical support of the Centre Technique Forestier Tropical (CTFT). The advantage of this project, compared to previous experiments, was that it allowed accurate measurement of the impact of silvicultural operations.

The project covers 1,200 ha and includes three field stations characteristic of the three ecological areas of the TMF of Côte d'Ivoire: semi-deciduous forest, evergreen forest and transition forest. Exploitation of the forest has been by traditional methods, followed by thinning by poison girdling. Two thinning regimes modelled on experience gained in Peninsular Malaysia were tested (30 and 45 per cent of the total basal area) by beginning systematically with the tallest trees in residual forest until the desired percentage of basal area was reached. The objective of the thinnings was to favour the valuable trees of more than 10 cm diameter breast height (dbh). No treatment has been planned to provide for regeneration by natural seeding but preliminary investigations seem to indicate that the natural regeneration induced is adequate for replacement.

After four years of observation, volume increment was 3.0 to 3.5 cu. metres/ha/yr compared with 2 cu. metres/ha/yr in control stands – a gain in growth of from 50 to 75 per cent based on measurements of stems over 10 cm dbh of seventy-three main species. Measurements taken every year showed that gain in volume increment increases with time and it is thought that the effect of the thinning operations will probably be felt for at least ten years. The largest yearly diameter increases, averaging 1 cm/yr, were found in species such as *Triplochiton scleroxylon, Terminalia superba and Tarrietia utilis.*

The economics of the operation are also promising, with 1 cu. metre produced for every US$5.6 invested, compared with US$7.4 per cu. metre for plantation-grown timber. If a wider range of species from the natural forest can be introduced in local markets, the economics of the operation would improve. The 10,000 ha Yapo forest will be managed experimentally on the basis of the results of this research project.

Although the period of the trial is still short, it seems fair to conclude that operations involving releasing of advance growth are by now well established but that methods of bringing about adequate natural regeneration of the desired timber species have not yet been mastered. Uncertainty also plays a role, as even management systems with a long research history such as TSS are only provisional, in that they have not yet gone through a full rotation of sixty to ninety years.

CONCLUSIONS

One can safely say that, at present, there are no sustained-yield management systems that have been practised for any length of time over any large area of forest in Cameroon, the Congo, Côte d'Ivoire, Gabon, Ghana or Liberia. On the contrary, tropical moist forest (TMF) is being destroyed at an alarming rate by unsustainable farming practices, with roads intended for logging or mineral extraction providing access to farmers. In Ghana and Côte d'Ivoire, more than 80 per cent of the original TMF cover has already disappeared, and there is increasing anxiety about the disruption of timber supplies as well as about the serious environmental consequences that might result with present trends continuing.

In Africa, natural forest management was practised in Nigeria, Ghana, Gabon and Côte d'Ivoire but it has been abandoned in favour of artificial regeneration methods. However, plantations do not constitute an alternative to forest management, because they do not yield the same environmental and conservation benefits or the same variety of non-timber forest products as managed natural forest; and even the timber produced is often technologically inferior due to faster growth. Plantations can play an important complementary role, but managed natural forest will have to provide an important share of future tropical timber supplies if these are to be sustained.

Forest management has been defined as "taking a firm decision about the future of the forest, applying it, and monitoring the application". The term has often been mistakenly used to refer to intensive management only. It can, however, be practised at various levels of intensity. As a minimum, it includes demarcation and protection, inventory, and the regulation and control of exploitation of the forest. More intensive management involves silvicultural interventions such as releasing and climber-cutting operations.

The reasons why there is so little natural-forest management are to be found in the context in which forestry is conducted: forest and land-use policy, forest legislation and reservation, the control of logging, socioeconomic and financial policy, taxation and economic circumstances. In the African TMF zone, most of the forested land is in the public domain, or at least nominally under government control. Good forest management depends, therefore, on the effective implementation of appropriate government policies. But, although most of the countries studied have committed themselves more or less strongly to sustained-yield policies in their forest legislation, little of this commitment can be traced in actual government policy.

More often than not, agriculture is perceived as more important than forestry and even isolated from it; and forests are seen as a convertible rather than a renewable resource. As a consequence, the forest legislation is not applied and forest protection not enforced. Whereas TMF cover in the six countries studied is estimated at 67.4 million ha, only 6.9 million – of which only some 4 million are actually forested! – have been reserved as production and protection forests. Furthermore, government forestry departments have been starved of funds, a fact which has seriously compromised their ability to apply the forest legislation.

The control of logging is equally problematic. In TMF in Africa, heterogeneity and low commercial timber volume effectively preclude the use of mono-cyclical (uniform) management systems; future timber harvests depend, therefore, on advance growth of commercial species left undamaged after logging. But detailed instructions to limit felling damage, as regards climber cutting, directional felling and marking of residuals, for example, are not mentioned in most of the present concession agreements. Furthermore, the absence of control or the unlikelihood of its enforcement does not encourage the loggers to comply with existing rules and regulations.

Socioeconomic and financial policies intended to improve forest utilization and prevent wastage have been introduced, but often without the desired effect. Efforts to promote the so-called lesser-known species (LKS) have had occasional success, but in most countries one to three well-established species still provide 50 to 80 per cent of the total log volume harvested. Policies intended to create employment by increased in-country processing have caused tremendous wastage of timber and enormous losses in government tax benefits forgone. Taxation policies have generally failed to improve the loggers' behaviour, as taxes have been too low and too little differentiated to be effective. Benefits allowed to national enterprises have led to inefficient forest exploitation, as they were not accompanied by appropriate credit and training facilities.

Economic circumstances, whether within or outside government control (indebtedness, world timber market fluctuations and uncertainties, the existence of inflation-prone non-exchangeable currencies, the lack of credit facilities), have often made a significant contribution to the factors constraining forest management.

Land-use and forest policy, and economic and financial policy, need to be improved to encourage sustained-yield forest management. The quality of these policies depends on the availability of good information, but research does not seem to have high priority in any of the countries studied.

This picture of the present status of forest management and its social, economic and political environment in Africa looks rather bleak. Admittedly, various measures have been taken recently, and it seems that governments are finally beginning to perceive the TMF as a renewable resource that should be managed in order to assure its continued existence and productivity. In Côte d'Ivoire there is firmer forest policy and good forest management research; Congo has adopted forest legislation conducive to sustained-yield forest management; Cameroon, a more efficient logging control system; Liberia, higher forest fees that are being reinvested in the forestry sector.

But, for sustained-yield management to be practicable on a large scale in the African humid tropics, all of those measures have to be taken simultaneously. If any one of the conditions discussed in this chapter is not fulfilled, then the tropical moist forest will continue to be pillaged for short-term gain, rather than managed for long-term, sustained benefits.

NOTES

1. FAO, *Management of Tropical Moist Forests in Africa*, FAO forestry paper (Rome: FAO, in press).
2. Grainger, Alan, *Tropform: a Model of Future Tropical Timber Hardwood Supplies*. Proceedings of the CINTRAFOR Symposium in Forest Sector and Trade Models (Seattle: University of Washington, 1987).
3. Schmidt, R.S., *Tropical Rain Forest Management: a Status Report* (Rome: FAO, 1987).
4. HIID, *The Case for Multiple Use Management of Tropical Hardwood Forests*, Study prepared for ITTO by the Harvard Institute for International Development (Cambridge, Mass.: HIID, 1988).
5. FAO, op. cit.
6. Repetto, Robert and Malcolm Gillis (eds), *Public Policy and Misuse of Forest Resources* (Cambridge: Cambridge University Press, 1988).
7. ibid.
8. FAO, op. cit.
9. ibid.
10. ibid.
11. Schmithüsen, Franz, *Forest Legislation in Selected African Countries*, FAO forestry paper no. 65 (Rome: FAO, 1986).
12. Catinot, René, *Etudes sur les systèmes d'aménagement dans les forêts tropicales mixtes d'Afrique francophone*, FAO forestry paper (Rome: FAO, forthcoming).
13. Repetto and Gillis, op. cit.
14. Anon., *A Global Research Strategy for Tropical Forestry*, Report of an international task force on forestry research (Washington DC: Rockefeller Foundation, UNDP, World Bank, FAO, 1988).
15. Burns, D., *Runway and Treadmill Deforestation*, IUCN/IIED Forestry Paper No. 2 (London: IUCN/IIED, 1986).
16. Lesser-known species are sometimes wrongly referred to as secondary species, although the fact that they are not extensively marketed is not necessarily related to the quality of their timber.
17. Repetto and Gillis, op. cit.
18. Anon., op. cit.

4. South America and the Caribbean
Timothy Synnott

This is an account of the present state of management of the tropical rain forest of the ITTO member countries in Central and South America and the Caribbean, namely Brazil, Bolivia, Ecuador, Honduras, Peru and Trinidad and Tobago. It is based mainly on information obtained during an eight-week tour in early 1988 followed by a workshop in Ecuador in July.[1]

The chapter describes attempts to initiate or maintain long-term yields and repeated harvesting of timber from natural or semi-natural tropical moist forest (TMF) and seasonal tropical broadleaved forests, and also describes silvicultural treatments, research, biological studies, inventories, and projects and plans which relate to these attempts. On the other hand, it excludes many other aspects of TMF and timber production, such as non-timber products (rattans, food, fuel, etc.), national parks and other areas of wildlife conservation, plantations and agro-forestry with planted trees, traditional woodsmanship and multipurpose resource management by forest people, and those logging activities (the majority) which are not part of a planned, sustainable programme.

The term "management" is sometimes used loosely in discussions. Some activities are called "management projects" which actually consist of demonstrations and trials. Equally, it is often said that management would be uneconomic when it is meant that silvicultural treatments would be. Here, management is understood to include many possible components: silviculture is one component of management, but it is possible to have effective management with few or no silvicultural interventions, and with only a few of the more important management activities.

Here we distinguish between the characteristic tools of *silviculture* and of *management*. Among the former we include such activities as the regulation of shade and canopy opening, treatments to promote valued individuals and species and to reduce unwanted trees, climber cutting, "refining", poisoning, enrichment and selection. Among the latter, the setting of management objectives, yield control, protection, working plans, felling cycles, concessions, roads, buildings, boundaries, sample plots, prediction, costings, annual records and the organization of the silvicultural work.

In many cases, it was not possible in the time available to obtain reliable figures for many aspects, such as the areas of forest in various categories, the analysis of timber production or the details of work done in particular places. It was often necessary to use information provided verbally from responsible sources or data from a single official report. It was often impossible to verify this information. When several sources of data were available, they were sometimes contradictory. Some figures, therefore, may be imprecise or out-of-date. Later or more accurate data may sometimes be available from other sources.

In recent years, many of the separate components of natural-forest management have been carried out in all member countries of ITTO in tropical America. These include the legal declaration of "forest reserves", inventories, silviculture, botanical and ecological studies, the issuing of harvesting licences and award of concessions with limits of size and yield, and some management plans. Further, national forest policies and laws all embrace (with many local variations) the concept of productive and sustainable management.

In all countries the forestry services have been involved in a wide variety of these management-related practices in different areas, often with foreign technical assistance, grants and loans in fixed-term projects. They also have some control over logging activities, by means of licences, concessions and roadside checks of log transport.

All tropical forests which have not been alienated to private ownership are legally state property, and available for allocation to one or other of the responsible government agencies, including the forestry service. Forest reserves for productive forestry, as well as national parks and other areas of conserved TMF, exist in all member countries.

However, from the viewpoint of professional forestry, this author has not identified any case of operational TMF management for sustainable timber production in any member country except Trinidad and Tobago. Even in Trinidad, management is not intensive but it does qualify as sustainable, even though silvicultural treatments are rarely applied and working plan prescriptions are not followed in strict detail. In other countries, in spite of striking advances during the past ten years or more, the following components are generally weak or lacking: advance planning of the location and intensity of the annual cut; supervision and control to ensure that the cutting conforms to the planning; and protection of the area to limit unplanned activities including settlement and uncontrolled logging.

Even in fully licensed concessions, forestry service control over logging intensities, the payment of fees and the implementation of rules concerning regeneration are far from complete. Over many forest areas, there are, in effect, no controls at all.

Nevertheless, many foresters and political leaders, most Indian and other communities, and some timber industrialists are becoming increasingly concerned about future timber supplies and environmental destruction. Some of the larger and more heavily capitalized sawmill and plywood companies already manage their harvesting with much more efficiency than most, and some forestry districts (e.g. von Humboldt, Peru) have recently increased their level of field checks. Honduras is implementing models of management for pine areas which may be extended to areas of TMF. It appears that a small increase in determination by certain governments and forestry services may shortly result in upgrading some concessions to the extent that they may validly be described as under sustained-yield management.

These favourable situations exist typically only where the forest resources have been seriously reduced, and where an expanding timber demand faces a declining resource. This does not yet apply over the bulk of Amazonia, where reservation and effective protection, and close controls over loggers, are not expected in the near future.

Trinidad is the only country of tropical America with a substantial history of professional management of natural forests, starting in the 1920s. It is the only country in which government foresters are carrying out natural forest management, on an operational scale, for timber production, over a substantial proportion of the national forest land. The managed area amounts to about 75,000 ha, which is most of the productive forest of the country. About 16,000 ha have been declared as fully regenerated after logging through the application of one of several silvicultural and management systems.

Most of these systems have involved control of logging intensities. The Trinidad Shelterwood System (which is no longer operational) involved more intensive utilization and some silvicultural treatment. The shelterwood system is an excellent example of a silvicultural system which was technically very successful in its time but which was discontinued when circumstances changed.

Honduras is the only one of the six ITTO member countries of tropical America which has natural pine forests. All forest management is the responsibility of COHDEFOR,[2] a semi-autonomous government agency. Unlike forest services in the other ITTO countries, COHDEFOR is self-financing from the proceeds of log sales. Even so, as in other countries, it has serious financial problems.

The continued conversion of TMF to grazing land, to permanent or temporary cultivation, or to wasteland is a process over which foresters alone have little control. Much of it is a spontaneous popular reaction to basic necessities or to forces applied from outside. In many countries, the allocation of land and credit are the responsibility of other agencies acting independently of the forestry service. Much settlement and forest clearance is encouraged by the practices and policies (either officially stated or unofficially implemented) of those in power. Only high-level political decisions will change this situation.

A forester with experience of practices in other tropical regions cannot help speculating as to why forest management has made so much less progress in South America than in other regions, which are often less developed economically. In many parts of Latin America, the early extractive, exploitative and not notably nationalistic practices and attitudes of the early landowners, of the old aristocracy and of the rich are still evident today. Many of those in power have supported the same interests or promoted the same practices (with conspicuous and often short-lived exceptions). In many areas, extractive enterprises still take precedence over national and social development. From these same principles are derived the land-use practices which are now eliminating the TMF: profit-oriented logging, enriching some but contributing little to state revenues or to local standards of living; and colonization and settlement, whether spontaneous (by landless peasants), privately directed (by aspiring landowners) or officially promoted (when it is described as land reform).

Certain clear conclusions were reached as a result of the study. One is that substantial quantities of commercial timber will in future be available from secondary forests, often privately owned or occupied, for which management techniques are not readily available. Another is that the prospects for sustained-yield management in reserved forests will depend upon the decision by governments, taken in the national interest, to exert more controls over commercial logging, to plan and co-ordinate the work of the various agencies concerned with agriculture, settlement and natural resources, and to give a higher priority to the needs of those living in and near the forest, especially Indian communities, small farmers and settlers.

FOREST LAND AND ITS CONTROL

In all member countries, forest land belongs to the state unless legally alienated to individuals or companies. In most countries the state also retains some legal controls over forest trees even on private land. The great majority of land covered with TMF is still state property, even though much temperate and subtropical forest has been taken into private ownership (e.g. a high proportion of the forests of southern Brazil and of the pine forests of Honduras).

But government ownership does not necessarily mean government management. Only a minority of TMF has been formally reserved as national forests or forest reserves (FR) and allocated to the forestry services for management, whether for production or protection. Even in the reserves, the forestry services usually do not have the resources to do more than maintain a presence. In other forest areas, though nominally state property, the legal responsibilities of forestry services are ill-defined and their resources even more reduced. Thus the great majority of the forests of the Brazilian Amazon are state forests but are not formally allocated to forestry.

All member countries have significant areas of TMF reserved in conservation areas such as national parks and forest reserves under a variety of names. Forest reservation is well established in Trinidad, where many reserves were declared and demarcated between about 1920 and 1940. Ecuador, Bolivia and Peru all have several reserved forests and, since 1985, Ecuador has had a comprehensive programme of defining and demarcating (but not yet protecting or managing) the reserves of Patrimonia Forestal, which are intended to be managed for a wide variety of purposes. Several recent attempts have been made in Honduras, too, to increase the area of forest reserves in TMF areas. In Brazilian Amazonia, the proportion of the TMF in reserves dedicated to forestry (Florestas Nacionais) is still minute.

Unfortunately, the nominal extent of reserved forests and conservation areas does not always correspond to reality, since many have been affected by settlements. Occupation, clearance, cultivation or grazing of forest lands leads, in all countries, to the forestry service losing what little token control it had over the land, whatever its legal status; unauthorized settlers are rarely expelled by the forestry services. On the other hand, there are cases where settlers and long-established Indian communities are driven out by later arrivals who have obtained title deeds or other means of control.

A summary of the forest areas and reserves in the region is given in Table 4.1. Details of the individual countries follow.

Table 4.1: Forest areas and reserves in the ITTO member countries of tropical America

Country	TMF area (ha)	TMF Reserves (ha)	%
Trinidad & Tobago	272,800	126,700	45.0
Brazil, Amazonia	315,000,000	> 800,000	0.3
Bolivia	45,000,000	10,000,000	22.0
Peru	70,000,000	52,000,000 reserved	75.0
		5,513,000 National For.	8.0
Ecuador	12,000,000	3,000,000	25.0
Honduras	2,500,000	250,000	10.0

Note
TMF Reserve areas include productive and protective forests, but not conservation areas. The table is not definitive, and is intended only for approximate comparisons, since legal categories and the accuracy of information vary greatly between countries.

Trinidad and Tobago

The areas of forest reserves are shown in Table 4.2. An analysis of aerial photographs, carried out during the 1978-80 forest inventory, covering most of the country, showed the following forest types: edaphic swamp forest – 16,400 ha; montane forest – 22,500 ha; evergreen seasonal forest – 115,200 ha; semi-evergreen seasonal forest – 14,100 ha; dry evergreen seasonal forest – 500 ha; deciduous seasonal forest – 3,700 ha; plantations, including non-timber – 21,400 ha; and secondary forest – 6,000 ha; totalling 199,900 ha.

Forest reserves are owned by the government and administered by the Forestry Division of the Ministry of Food Production, Marine Exploitation, Forestry and the Environment. This division includes a section for national parks and conservation, and the Forestry Research, Inventory and Management section (FRIM) which has been responsible since 1978 for inventories, working plans, tree marking and research. Forest lands are liable to be affected by the State Lands Project, which is responsible for identifying areas of state-owned lands suitable for agriculture and allocation to farmers. Between 1979 and 1982 many farmers left the land but, more recently, there has been an increasing occupation of forest lands. Other state-owned forest lands are assigned to other ministries, including water authorities and the military. In addition, forestry activities are increasingly affected by public opinion and by organizations such as the Field Naturalists Club.

Table 4.2: Areas of land and of forest in Trinidad and Tobago

Total land area (ha)	Trinidad	476,900
	Tobago	35,500
	Total	**512,400**
Forest land (ha)	Private	54,400
	State-owned	218,400
	Total	**272,800**
Forest Reserves (FRs) (ha)	Production	94,500
	Protection	32,200
	Total	**126,700**

Note
Production FRs classified as:
"intensively managed plantations"	17,500
"intensively managed natural forests"	16,000

Not all forest reserves are available or intended for long-term timber production. Some are dedicated to wildlife conservation, and others are too steep or inaccessible to be commercially harvested. But about 75,000 ha of natural forest are now classified as "productive" (intended for long-term, sustainable timber production), of which 16,000 have been classified as "intensively managed". It is justifiable, however, to describe the whole forest reserve system as "managed" and, for ITTO purposes, the 75,000 ha of natural "production" forest as "managed for sustainable timber production". They are protected to a degree by resident forest guards, their objectives of management have been defined, most of them are covered by working plans (albeit due for revision and not fully implemented), and logging is subject to some control. The 16,000 ha classified as "intensively managed natural forest" have been logged and then closed until the next cycle, according to one of the silvicultural and management systems described below (pp.93-4).

Table 4.3: Forest areas in Brazil

Total land area		851,196,500 ha
of which: Legal Amazon		500,000,000 ..
of which: Brazilian Amazonian forest:		
Closed dryland forest		180,000,000 ..
Open forest		76,600,000 ..
Transitional, seasonal forest		42,300,000 ..
Dry forest		10,000,000 ..
Riverine, swamp and mangrove forest		6,000,000 ..
	Total: approx.	315,000,000 ..
including:		
Areas of logged and partially cleared forest		13,500,000 ..
Conservation areas	about	12,000,000 ..

Brazil

Forest statistics are shown in Table 4.3. Estimates of forest land differ widely depending on the definition used and on allowances made for clearances.[3] Trustworthy figures for forest ownership hardly exist. FAO reported that "public" ownership of existing forests varied from 93 per cent in the northern region, where most of the Amazonian forest occurs, to 70–87 per cent in other forest regions (but only 29 per cent in the south). In Acre and Amazonas, almost all the forest land is legally federal property, although large areas are "occupied". In Pará perhaps half the forests in the north are federal property but few in the south. Most of the Amazonian forest technically belongs to the federal government, and responsibility for the timber was vested in IBDF, which also had responsibility for the National Forests (Florestas Nacionais), National Parks and Biological Reserves. IBDF has recently been disbanded and most of its functions have been taken by the new Brazilian Institute for the Environment and Renewable Natural Resources.

There has sometimes been opposition to forest reservation from other groups with an interest in forest resources and land, including FUNAI (responsible for Indian community affairs), Petrobras (the oil industry, with interests in access for exploration and exploitation) and other private concerns, who suggest that forest reservation prejudices the potential development of the land. Progress in forest reservation has been slow; in fact, reservation for biological conservation has been far more extensive, but parks and reserves have also been affected by other interventions.

Bolivia

Forest statistics are given in Table 4.4. Much of the vegetation, especially in the highlands, has been affected by farming, grazing, fire and logging. In 1978, the existing forest cover was estimated at about 55 million ha (50.8 per cent of the land area). The TMF areas of the Amazon basin (*selva alta* and *selva baja*) have been estimated as shown in Table 4.4.[4]

Table 4.4: Forest areas in the Bolivian Amazon

Total land area	109,850,000 ha
Productive TMF intact	17,250,000 ..
Productive TMF logged	12,220,000 ..
Total productive TMF	**29,470,000 ..**
Unproductive TMF (inaccessible etc.)	14,100,000 ..
Total TMF	**43,570,000 ..**
Barbecho, scrub forest	1,400,000 ..
Total forest in Amazon basin, approx.	**45,000,000 ..**

All natural forests belong to the state, whatever the status of land ownership. Legal responsibilities are divided. The Centro de Desarrollo Forestal (CDF) has technical and administrative responsibilities for forest resources. The Instituto Nacional de Colonisación (INC) directs and plans new settlements. The Consejo Nacional de Reforma Agraria (CONRA) is responsible for allocating land for agriculture. Despite legal provision for a joint Comisión del Uso de la Tierra, there is in practice little co-ordination between these three. Loans to farmers are conditional upon land improvement, which specifically includes forest clearance, usually without any consultation with the forestry authorities.

The laws provide for several different kinds of classified forests and conservation areas, in which the land and forest cover have special legal status. In total, they amount to about 12 million ha. The most important categories are Bosques Permanentes de Producción (BP) and Reservas Forestales de Inmovilización (RFI). Included is Los Chimanes, Beni, an area of 1,100,000 ha which was declared an RFI in 1978 and a BP in 1986. This is one of very few reserved forests relatively free of colonization by farmers which has been declared a BP.

Twenty conservation areas have been declared with many different designations (including Parque Nacional, Reserva de Fauna Andina, Refugio de Vida Silvestre, Reserva Fiscal, Reserva Nacional and Santuario de Vida Silvestre). Many of these areas contain TMF. In practice, the distinction between the categories is unclear, protection is minimal, management non-existent and occupation by farmers locally serious. Conservation areas in TMF include the two national parks, Prof. Noel Kempff Mercado (*c*.900,000 ha), which is the only one demarcated, and Amboro (67,000 ha).

Table 4.5: Forest areas in Peru

Total land area	128,521,600 ha
of which forest areas:	
TMF	67,700,000 ..
Dry tropical forest	525,000 ..
Swamp forest	1,050,000 ..
Mangrove	28,000 ..
Montane forest	7,000 ..
Podocarpus	370,000 ..
Total	**69,680,000 ..**

Note
The TFAP estimated a total forest area of 77,600,000 ha including 73,700,000 ha of TMF of which 45,000,000 ha are potentially suitable for managed production forestry.
Source: FAO, *Los recorsos forestales de la America tropical* (Rome: FAO, 1981).

Peru

The forest statistics are given in Table 4.5. All natural resources, renewable and non-renewable, including forests, are national property. Forest areas have special legal status, are the responsibility of foresty authorities and are subject to forest laws. The following are the approximate areas: Bosques de Libre Disponibilidad – 36,700,000 ha; Bosques de Proteccion – 4,342,000 ha; Bosques Nacionales (5) – 5,514,000 ha; Unidades de Conservacion (22) (national parks, natural reserves, natural sanctuaries, historical sanctuaries) 5,500,000 ha. The total reserved area thus amounts to 52,056,000 ha.

The Bosques Nacionales (national forests) were reserved during the 1950s and 1960s. The actual forested areas are less than the totals given, due to conversion to agriculture, and areas of steep slope or high altitude. The area of national forests available for productive forest management is also limited by security restrictions and proposed conversion to national parks; it totals 2–3 million ha. The Bosques de Libre Disponibilidad, on the other hand, were declared in order to facilitate the issue of logging licences, and were not originally intended for long-term sustained-yield management. They were not based on detailed studies of land-use potential; they include many areas with steep slopes and exclude some good productive forests.

Table 4.6: Forest areas in Ecuador

Total natural forests	about	12,000,000 ha
of which:		
Andean, montane		1,000,000 ..
Coastal TMF		2,000,000 ..
Eastern Amazonian TMF		9,000,000 ..
Mangrove		150,000 ..

Ecuador

Forest areas are as shown in Table 4.6. There is substantial uncertainty about the areas of forest, since there has been no recent national inventory. All forest land legally belongs to the state unless it has been assigned to an individual or organization. A few timber industries own forests amounting to several thousand hectares. In addition, farmers, cattle raisers, communities and colonists occupy forest land in varying stages of deforestation.

Of the forest land which is still unoccupied and legally in state ownership, much has no special legal status. The following categories of conservation or protection status are recognized under the forest law: Patrimonio Forestal del Estado (permanent forest estate); Patrimonio de Areas Naturales del Estado (conservation areas); Bosques Protectores (protective forests). These reserved areas total

6,104,841 ha. They are almost exclusive of each other, but there is some overlap. Some protective forests occur in conservation areas, and it is expected that future Patrimonio Forestal areas will include many areas which are now protective forests. Further, all categories include non-forest areas; some have been recently cleared of forest, but others, such as high montane areas and parts of Galapagos National Park, have not carried forest in recent times. DINAF has a degree of responsibility for all these categories, but some protective forests are actually privately owned.

A programme for reservation of a permanent forest estate started in the early 1980s. Two provinces have been completed and the results for four others are expected in 1989. It is the work of an interdisciplinary commission including DINAF and IERAC. It involves the definition of those areas, mainly natural forest and mainly unoccupied by farmers or indigenous Indian communities, which are suitable for long-term management. Results so far indicate 21 per cent of Esmeraldas province and 30 per cent of Napo. This is a more developed programme of forest reservation than any in the neighbouring countries, although there is at present no finance, staffing or infrastructure for the maintenance or protection of the declared areas. The forest laws provide for a wide variety of conservation areas, but the differences in their status and objectives are not always clear or defined. They total about 2.9 million ha including, among the six national parks, the Galapagos and, in TMF, Yasuni. The protective forests include all present, former and potential forest lands which are of special importance for protection of soil or water resources, or for preventing erosion or flooding. They amount to about 1,350,000 ha and are likely to be increased to 2 million ha. In practice the long-term status of these areas is still at risk, due to the shortage of staff and funds, and the divided responsibilities within government. While DINAF is responsible for forest land and resources, INERH is responsible for settlement, colonization and watersheds.

Honduras

No nation-wide inventory has been carried out recently, only selected areas having been covered. Estimates of the decline in area of TMF vary between 50,000 and 80,000 ha a year. Approximate estimates of forest cover (for international comparison only) are given in Table 4.7.

Many forest areas have been decreed as Reserved Forests, following the provisions of the forest laws, but no TMF areas have yet been declared as production forest reserves for sustained timber production, and none is managed as such, although a new management plan is now being put into operation in a TMF concession, TIMSA. The only production forest reserve, Olancho (total area 1,437,000 ha; declared in 1966), is mainly pine forest. However, Olancho has a substantial area of unmanaged TMF, and there are many other areas with TMF reserved for conservation, environmental protection or research. These include eleven Parques Nacionales (among them the Rio Platano biosphere reserve), eighteen Reservas Ecológicas and eight Refugios de Vida Silvestre. Almost all cloud forests (those at high altitude that receive their precipitation from cloud or mist) are protected.

Table 4.7: Forest areas in Honduras

Total land area	11,208,800 ha
Forest areas:	
Closed forest	4,200,000 ..
Scrub (barbecho)	1,000,000 ..
of which:	
Closed broadleaved forest	2,000,000 ..
Open or degraded broadleaved forest	500,000 ..
Broadleaved scrub	500,000 ..
Pine forest	1,500,000 ..
Open pine stands and scrub	500,000 ..
Mangrove forest	145,000 ..

The legal division of responsibilities for national parks, ecological reserves and wildlife refuges is not entirely clear. The forest law provides for these categories but without defining them in detail, while other laws imply responsibility divided between COHDEFOR and DIGERENARE. Neither organization has sufficient funds to ensure protection and management.

On average, 60 to 80 per cent of forest land is owned by the state and, even on legally registered private land, the commercial harvesting of trees is, in principle, subject to controls by COHDEFOR.

FOREST LAWS AND POLICY

All member countries have comprehensive bodies of forestry legislation and official, if general, expressions of national forestry policy, but, also in all countries, it is considered that parts of the laws are out of date or no longer applicable; and revisions are under way or proposed.

In spite of many similarities in the laws, there are some important differences. A notable feature is the legal status of COHDEFOR, the forestry service of Honduras, which was formed as a self-financing para-statal organization in 1974 with the President of the Republic as its president.

The results are evident in the realistically high fees charged for standing timber, the relatively generous staffing and the new management initiatives being developed. Another difference is in the relative degrees of responsibility for forest management and regeneration of the forest services. In Brazil, Peru and Bolivia the concession holders have a legal responsibility to carry out forest regeneration, although in many cases these are evaded or else fulfilled by paying a small regeneration fee to the forestry service. The smaller logging companies are not expected to carry out regeneration, which has resulted in a proliferation of small operations, notably in Peru and Ecuador. In Honduras, COHDEFOR retains almost full responsibility for at least the supervision of management operations in the natural pine forests. The few cases of post-logging regeneration activities occur either in the state-managed forests in Trinidad, or in isolated projects elsewhere, or in privately owned company forests such as Jarí. A detailed comparative study of national forestry laws would yield valuable conclusions for governments wishing to improve their management. However, the most important feature of forestry legislation in these countries is the minimal level to which the laws are respected or enforced; this is common to all countries.

Trinidad

The Forest Act dates back to 1915 and, in 1942, a forest policy was accepted by government and is still officially in force. In 1980, the Forest Department drafted a new forest resources policy document with two main aims: to allocate an adequate area of land in strategic places for forestry purposes; and to manage these resources for optimum combinations of their productive, protective, recreational, aesthetic, scientific and educational capabilities. This policy has not yet been officially adopted by the government but it serves as a guide for forestry activities within the limitations imposed by staff and budgets. A national plan for agriculture, forestry and fisheries and a national forestry conceptual plan are also being prepared.

Brazil

A forest law (Codigo Florestal) was enacted in 1939. This was followed by a new law in 1965 (Decree Law No. 4771 of 15 September 1965) which fully covers forest ownership and exploitation and imposes regulations concerning fees and regeneration after logging. Among other provisions, Article 44 requires that 50 per cent of all new agricultural holdings in forest areas should be maintained under the natural forest. Timber fees and other charges are based on this law but the full provisions are not always followed.

A new law was formulated in 1986 modernizing many of these provisions and increasing the legal responsibilities of private forest owners and loggers for managing the forests. There are also requirements that loggers should plant a prescribed number of trees for each cubic metre of timber extracted. There is some uncertainty whether this law has been fully enacted but certainly it is not yet being followed or enforced to any significant extent. However, a few companies are reported to have written management plans and undertaken modest replanting programmes.

Bolivia

There is no separate declaration of a national forest policy but two major forestry laws, one establishing CDF, the other defining rules governing harvesting, licences and concessions and the responsibilities of concession holders for reforestation. They provide for creation of a National Forestry Fund and Land Use Commission, but these provisions have not yet been implemented. Since 1986 the responsibility of CDF has been decentralized.

Peru

Forestry is governed by a large number of laws and decrees (Decreto Ley, Decreto Legislativo), often supplemented by *regulamentos* (regulations) which give more technical details and instructions.

There is no separate forest policy document. The Decreto-Ley 21147, Ley Forestal y de Fauna Silvestre, of 13 May 1975 prescribes the categories of reserved forest land: Bosques de Libre Disponibilidad (BLD) – for permanent production of forest products and wildlife (now totalling over 36 million ha); Bosques Nacionales (BN) – reserved in the 1960s for use and management by the state, but now available for logging contracts (about 5.5 million ha); Bosques de Protección (BP) – mainly for protective functions; and Unidades de Conservación (UC) – totalling 5.5 million ha. Provision was included for timber harvesting and that logging should be regulated by management plans. Decreto Legislativo 2, Ley de Promoción y Desarrollo Agrario (1980), stipulated that agriculture and livestock production will have priority in suitable areas of the forest regions but the conflict between forest protection and colonization has continued.

A proposed new law, Código del Medio Ambiente, will, if approved, result in a major change in the concession system, giving the DGFF much more power over the selection and delimitation of annual logging areas. Studies are also under way to revise and collate the forest laws and regulations, and to prepare a Texto Unico Concordado del Ley Forestal. The newly prepared Tropical Forest Action Plan includes a recommended policy for forestry and fauna.

Until recently, administration and control were in the hands of DGFF, while research was carried out by INFOR as a separate dependency of the ministry. Since 1987, these two arms have again been combined within DGFF. In the provinces, management (*actividades normativas*) and research were based in the Centros Forestales (CENFOR); these are now being separated, as part of a process of decentralization. The management and supervision of Bosques Nacionales will remain under the direct control of DGFF.

Ecuador

The Forest Service, originally PRONAF, now DINAF, was established by law in 1952. The present forest law, Ley Forestal and Ley de Conservación de Areas Naturales y Vida Silvestre, dates from 24 August 1981. The Forest Regulations date from 1983. These laws include provisions for declaration of reserved forests, including the Bosques Protectores, the Patrimonio Forestal and the Patrimonio de Areas Naturales. They also include regulations for logging licences and concessions. The principal deviations between reality and the implied policy and laws are the absence of management in the former concessions and in most of the existing reserved areas.

Honduras

The national forest policy is stated in the forest law, Decree 85 of 1971, Chapter 1, Generalidades. Both policy and law deal comprehensively with the production systems, conservation and protection, multiple use and national welfare. In 1974 COHDEFOR was established as the national forestry agency, semi-autonomous and self-funding, responsible for protected forest areas, declared by decree, and for the sale of standing trees. Until recently, COHDEFOR was also responsible for the sale and export of sawn timber, but sawmills and other companies now market their own wood products.

The Reglamento General Forestal (1984) gives detailed technical regulations for reservation, protection and uses of forest land and products. A revised forest policy has been prepared by COHDEFOR, and was approved by the President in September 1986, but it has not yet been fully put into practice. Other laws also affect forest land, in particular the Law of Agrarian Reform, by which INIA has legal custody of the land while COHDEFOR has custody of the trees.

FOREST MANAGEMENT AND SILVICULTURE

The only known cases of operational TMF management for sustained yields of timber are in the Production Forest Reserves of Trinidad, covering about 75,000 ha. The management there is not intensive and silvicultural principles are implemented mainly through logging controls, but sufficient criteria are satisfied for this to be called sustained-yield management. As indicated above, the forests have a degree of protection; their objectives of management have been defined; most are covered by working plans (although not up-to-date or fully implemented); existing logging is subject to some control and planning; and measurements and experience show that timber production is sustained.

Elsewhere, major attempts at integrated forest management, with full management plans, have been initiated by forestry services only in Tapajós National Forest, Brazil, and in von Humboldt National Forest, Peru, both with FAO assistance. In neither case were the project proposals fulfilled or operational management implemented, for reasons which included the market demand situation in Tapajós and the high investment proposed in von Humboldt. However, Tapajós is still intact; and in von Humboldt pilot-scale regeneration has covered nearly 1,000 ha, and the forestry service is gradually exerting more control over the logging concessions. In Chimanes National Forest, Bolivia, a new integrated management programme is now starting, involving several organizations and with more realistic but still ambitious objectives. In Honduras, a new management plan for El Zapote and other initiatives are under way.

These are the most serious management attempts so far, but there are many other management-related activities carried out in forestry service programmes, research areas and technical assistance projects. These certainly have a local or regional effect on the forests and on production, but do not, in the opinion of this author, amount to operational sustained-yield management.

Nevertheless, this does not mean that timber production will not be sustained into the foreseeable future. Great areas of forest remain unlogged. These are steadily being affected by disturbances; much will be permanently converted to other land uses, but much will remain, after selective logging or temporary clearance, as disturbed and secondary forest. There are already several million hectares of these kinds of forest, now unmanaged but growing and regenerating, which will continue to yield timber. Also, many private or occupied forest areas will provide a sustainable yield, although not necessarily as a result of conscious planning or management. This yield will doubtless include a different mix of species and volumes from the first logging of primary forest, but it will provide plenty of timber usable in the less selective markets of the future. This will occur even if cultivation is repeated occasionally, and especially if some agroforestry practices are introduced.

In some concessions (e.g. around Belém, Brazil; Santa Cruz, Bolivia; Pucallpa, Peru) logging is being done with increasing efficiency, and owners are concerned about future sources of timber. Locally, the staff and resources of the forestry services, and the awareness of political leaders, are increasing. In these areas, where growing industrial demand faces declining resources, there are good prospects for integrated forest management to be started in the near future.

Many timber inventories have been carried out in the TMF areas of member countries but there is doubt about the reliability of some.

More important is the general conclusion that inventory results have rarely been used by forest services or industries for detailed planning of logging licences or operations.

Trinidad

Largely due to its colonial past and the efforts of some innovative individual foresters, Trinidad has a long history of forest reservation, protection, management and silviculture very different from that of other countries of tropical America. The professional foresters were mainly members of the colonial civil service, and as such enjoyed many advantages. They had relatively easy access to the technical literature on tropical forestry, written mainly in English; they frequently had personal experience of practices in other tropical and subtropical countries; there were incentives for professional advancement through technical achievements which were regularly publicized through quarterly, annual and technical reports; and they had access to specialist technical advice from the colonial forest service and the then Imperial Forestry Institute, and from foresters with years of practical and operational, as well as academic and research, experience in the tropics.

These factors ensured that many forestry practices in Trinidad, such as the demarcation, gazetting and protection of forest reserves, and even the forest laws and policies, had much in common with those in countries with larger forestry departments and resources.

A 2 per cent inventory conducted between 1928 and 1938 covered most of the present production forest areas. The results were used as the basis for management control of logging and preparation of working plans, while between 1978 and 1980 a comprehensive national forest inventory of the natural forests was carried out using strip samples.

Continuous forest inventory using permanent sample plots (PSP) started in 1983 in the natural forests. These are the only PSP established in tropical forest in America as part of a statistically designed sampling programme for forest management rather than as a research study.

Working plans were written from 1935 onwards. They included limitations on the felling of Class 1 trees and aimed at achieving a sustained yield of timber, controlling logging by area or size limits coupled with natural or artificial regeneration. Subsequently, although management continued, the plans were not revised or management did not follow the detailed prescriptions. The most recent working plan was written for Arena Forest Reserve but the felling programme is not yet being carried out.

A number of well-developed systems of silviculture have been used in Trinidad.

The *open range* (or *selection*) *system* has been the system most widely followed since the start of professional management, and it still operates in some forest reserves (e.g. the Valencia Range). Individual licensed loggers are licensed to cut a specified volume or number of trees and may select them anywhere in the reserve or range as defined and approved by the Forest Department. In many reserves this has amounted to a "loggers' selection system" without close control by the department. In some relatively inaccessible areas (e.g. Victoria Mayaro FR) felling has not been heavy and the stocking of seedlings, saplings, and trees of all sizes remains adequate to maintain a continued cutting programme.

In areas of closer supervision, especially in the *Mora* forests (almost pure stands with abundant representation in all sizes), it has been closer to a true "selection system" with excellent results. It functioned as a selection system in areas where logging was restricted by a girth limit, and where replacement of valuable trees occurred within a defined distance. But, in areas of poor control and of high demand, heavy localized felling has resulted in degraded areas with inadequate young trees of valued species. *Mora* regeneration suffers from early exposure to full light and has been badly affected in some places. Many of the worst-affected areas have been converted to plantations. Other areas were overcut and will not be ready for the second felling cycle after twenty-five years, as planned.

The *open range system within blocks* is a later variant of the open range system. The forest is divided into blocks within which the loggers' selection continues in succession. Staff open and close each block to logging when appropriate. This system was introduced in 1948 in the *Mora* forests of north Trinidad and in 1954 in the south. Initially, especially in the Catshill Range, the block system was part of a programme of intensive logging and conversion of natural forests to plantations, but from about 1960 it became a periodic block system often performing as a true sustained-yield selection system with an intended cutting cycle of twenty-five years. It is still operational.

The *silvicultural marking within blocks* or *periodic block system with silvicultural marking* involves more intensive management and is a considerable silvicultural improvement over the first two.

Department staff mark all trees to be cut in order to maintain a satisfactory forest structure and distribution of young and/or seed-bearing trees of valued species. Not all the valuable trees of useful sizes are marked, but many less valuable trees are, to promote their sales and to maintain or improve the condition or species composition of the forest. This system was introduced in about 1974, e.g. in Victoria Mayaro, but has not been fully effective because the loggers were not persuaded to cut the less valued species. In 1986, 400 ha were marked for this system by FRIM.

Silvicultural marking in blocks with compulsory felling, or *periodic blocks with silvicultural marking and felling*, is the newest method, being applied in Rio Claro West Range. The licences will require the loggers to cut all marked trees. The tree marking is intended as a crown thinning to favour trees for the next cutting cycle in about twenty-five years, and thus improving stand composition and quality, while removing 8-10 trees (or 25 cu. metres) per hectare. If supervision and the incentives in the fee system are adequate, this method is expected to be successful both for silviculture and for management.

The *Arena Tropical Shelterwood System* or *improvement fellings* is a well known system which has resulted in valuable stands that are now ready for recutting, although the silvicultural treatments involved have not been applied for several years. In Arena FR, the silvicultural management of natural regeneration started in 1929. The system developed gradually with careful control of the logging intensity and canopy opening, tending and weeding of the regeneration, and removal of unwanted or defective trees for fuelwood and charcoal. It reached its peak in the 1950s, with operations spread over ten years, including climber cutting, partial removal of the canopy to form a shelterwood, tending of regeneration, and final removal of shelterwood.[5]

The success and cheapness of the system depended largely on the charcoal burners. They were allotted 0.4 ha at a time, and used all wood which had not been sold for timber, even partly rotted and fallen trees. Subsequently, after the regeneration was established, the upper canopy was gradually killed by poisoning and felled by the charcoal burners only when dead, with minimum damage to the young regeneration.

In the mid-1950s, the system was judged to be so successful that the cutting cycle was reduced from sixty to thirty years, a change from a mono-cyclic to a poly-cyclic system. Some areas have already received their second harvesting, and Arena FR is now ready for a full second felling cycle. The intensity and system of harvesting have not yet been finally decided.

The market for firewood and charcoal declined rapidly from the mid-1950s and more reliance has been placed on poisoning during the last years of the system in Arena. By about 1963, the whole of the appropriate sandy-soil types in Arena had been treated and regenerated and the system ceased to operate. The forests on the clay soils, especially the *Mora* forests, did not regenerate well under the open shelterwood canopy and were regenerated under a selection system or converted to plantations.

Some selective tending and weeding have been carried out in Arena in the 1980s but the system as a whole is no longer applied. It seems that about 2,500 ha were treated with the shelterwood system up to 1978. This was thought to be successful, but no formal research or measurements have been carried out to compare the shelterwood system with others.

The areas treated by this system now carry full stands composed almost entirely of sound individuals of valuable species, ready for a second felling. Quantitative information about volume production will soon be available from the many PSP in Arena and elsewhere. Average yields have not yet been calculated, but some estimates indicate a mean annual volume increment of 5 cu. metres/ha/yr. In 1957, the costs for the full 30-year cycle were calculated at 60 man-days per hectare with charcoal burning, or 87 man-days with only poisoning. This system was a conspicuous silvicultural and management success in its time. Among many reasons given for not applying it more widely, in addition to lack of money and post-independence shortages of trained staff, were the declining market for firewood and charcoal and the generally satisfactory results of the selection or periodic block systems when correctly applied.

It is expected that *future management* will be based on the selection or periodic block system with silvicultural marking and at least some compulsory felling of marked non-timber trees with a cutting cycle of 25–30 years. Silvicultural interventions in future will probably be mainly restricted to tree marking before felling and some cutting of climbers. Production is expected to be lower than the harvest which could be achieved in the intensive shelterwood system but more than the 1 cu. metre/ha/yr provisionally estimated for the open range system.

Brazil

There are no areas of TMF management for the sustainable production of timber, in the full professional sense; but there are several areas where distinct elements of forest management are being or have been carried out on a research, demonstration, pilot project or even operational scale. These elements include, in different areas, planned and programmed logging; written management plans; studies of forest ecology, phenology, regeneration and botany; application of silvicultural treatments including selective poisoning and thinning; enrichment and replanting trials; and repeated measurements in permanent plots.

In some areas, such as Curuá-Una, Tapajós and Jarí, these activities have been carried out for so long, or on such a scale, that they may be considered as part of an incomplete management system. In these and perhaps a few other areas, only a few more elements are needed to constitute a complete (if still simple) management system.

The first timber inventories in Amazonia were apparently carried out in 1951–3 with assistance from FAO. These covered the whole of Brazilian Amazonia, at a very low intensity, including some sampling lines near Santarém. Subsequent inventories have shown the regional variations in more detail. Between 1956 and 1961, a second inventory was carried out with FAO assistance in Pará. From 1968, an extensive survey, covering the whole of Amazonia, was carried out by Radam-Brasil, based mainly on radar-photos with some ground checks. Other more detailed inventories have been carried out by IBDF and other agencies over 1.8 million ha of TMF.

Tapajós National Forest has a full management plan. This is the only management plan for government-owned TMF in Amazonia, but it has not been possible to implement the prescriptions which relate to operational management. Many of the planned research activities are still being carried out, but protection and controlled harvesting have not yet been implemented as planned.

A number of logging enterprises in Matto Grosso, Pará and Rondonia are reported to have management plans, with some progress in implementing them. Some properties in Pará also have written management plans, but are not yet fully implementing them.

Silviculture and Management in Curuá-Una

The Curuá-Una Reserva Florestal was designated in 1958 with an area of 71,250 ha. It included an experimental reserve of 1,800 ha. Forestry work started in 1958 and experimental field work in 1959, as a result of an agreement between FAO and SUDAM. After FAO staff left, research and other activities were later resumed by CTM, Santarém, and FCAP, Belém. Substantial areas of species trials were planted, with some studies in the natural forest. J. Pitt established a TMF silviculture-treatment trial, the first in all Amazonia. The treatments included thinning, underbrushing and poisoning.[6]

In the 1970s, until 1978, a programme of controlled mechanized logging was carried out and forty permanent plots were established in 1978–9. The same areas are now expected to receive a controlled thinning in a new treatment trial starting in 1988, carried out by CTM. Current plans include a new management programme covering 5,000 ha. Inventories are now being carried out.

Silviculture and management in Tapajós and Belterra

Tapajós National Forest was founded in 1974. Field work started in about 1972, including FAO involvement from 1972 to 1982. The implementing agency working with FAO was initially IBDF-PRODEPEF and later EMBRAPA-IBDF. Trials started in 1975 with a timber inventory, linear regeneration sampling and some planted species trials, followed by more detailed inventories. In 1979, a controlled trial logging was carried out by SUDAM, at two intensities. Further diagnostic sampling was carried out in 1981 and 1986. Permanent plots were established in 1981, and remeasured four times to 1987. A phenological park of 400 ha was established in 1980 and a tree-girdling trial in 1986.

A new study of different logging intensities was started in 1981, incorporating lessons learned from the earlier trial. A 100 per cent pre-exploitation inventory was made over an area of 144 ha, and forty-eight permanent plots were established. A controlled logging was carried out in 1982, with two intensities. Plots were remeasured in 1983 and 1987. Early observations suggest that the actual impact of the controlled logging was highly variable without important differences between the two intended intensities.

In Tapajós, IBDF has made its first attempt at integrated forest management in TMF, to follow the prescriptions of the management plan.[7] A timber inventory of about 4,000 ha was carried out in 1983–4, and it was proposed to arrange for felling of 1,000 ha per year under licence. Logging areas in units of 100 ha were advertised. Several local logging companies applied, but only one accepted the IBDF conditions and carried out a commercial logging over 100 ha in July 1987. It is not yet known when licensed logging will continue. The lack of success in implementing the full planned cutting programme has been attributed to several factors. Local loggers obtain almost all their supplies from land newly occupied and cleared for agriculture, with no official restrictions and at minimum cost. The IBDF licence prescribed the full legal fees, and also the cutting of certain potentially usable (but not generally used) species. IBDF licences were not available for more than one year at a time, making it less attractive to invest in infrastructure. These factors, in a region of still readily available cheap logs, made the proposed logging less than attractive for the loggers. No treatments have been carried out.

At Belterra, part of the old Ford rubber plantation, the main emphasis has been on species trials, with about 80 ha of trial plantations established. In addition, there is a secondary forest study area of 132 ha, with twenty permanent plots. Studies are concentrated on natural regeneration, succession and phenology.

Silviculture and Management at Jarí

Jarí Cia Florestal Monte Dourado is now owned by a consortium of Brazilian business interests, including a government-owned share in the pulpmill. Jarí is not at present carrying out TMF management for long-term timber production. However, their forestry operations, including felling and extraction, road building and maintenance, mapping, research, and general control and monitoring, are outstanding examples of management. Moreover, Jarí has recently started to take a serious interest in systems of partial and selective exploitation of TMF and in monitoring their effects. The level of technical organization and control in Jarí is outstanding in Brazil, and it can be expected that their studies will reach practical conclusions relevant to management elsewhere.

The pulpmill uses wood from the natural forest as well as plantation wood. Jarí runs a 55 megawatt thermo-electric plant, requiring 2,000 tonnes of wood fuel daily. To supply this, 1,500 ha of natural forest are clear-felled annually and replanted for pulpwood. The wood extracted is used mainly for fuel, while some is used for pulp, posts and poles, and some of the best logs are converted to sawn timber in the company sawmill, for local use. Jarí foresters are studying ways of extracting sawlogs and/or smallwood without causing such complete forest clearance and soil exposure. These studies involve staff of EMBRAPA and of the research, extraction and production departments of Jarí.

The main formal trials are known as Felipe IV and Felipe VI. Their designs are based on two possible management systems: a complete clear-cut, followed by natural regrowth and regeneration, to be used for pulpwood, fuelwood or both (as successfully carried out in Colombia); and a selective logging for timber, with or without subsequent treatments. Interest is now also being focused on a shelterwood regeneration system, yielding both timber and pulp/fuelwood, because of the possible damage caused by complete clearance and restrictions which may be imposed in the future.

Operational clear-felling is itself a major enterprise, and is organized in a well-defined sequence according to the distance from the main road, the size of trees and the season. Extraction vehicles are of three sizes, from Brazilian-made mini-skidders to very large rubber-wheeled caterpillars. Natural forest is left uncut in valleys and on steep slopes. An operational trial is now being made of selective logging.

Silviculture and Management of Florestas Río Doce
This company is one of very few working TMF that has an interest in regeneration and long-term production. Their forestry interest is principally in charcoal, because of the present industrial demand and the much larger demand expected in future in Amazonian industrial development projects such as Carajas. Timber is expected to be an important additional product at the first cutting, and may be included as an objective of the regeneration systems. The company is operating in a number of areas, each with controlled cutting and monitoring of the regrowth: Jequitinhonha, Minas Gerais (for charcoal); Linhares, Espírito Santo (since 1980); Buriticupu, Maranhao; Carajás, Pará (planned); and Maraha, Pará (since 1974). The only area of operational, commercial production is in Minas Gerais. The other areas are said to be mainly for demonstration and research, to determine (and publicize) the sustainability of production.

The management role of EMBRAPA, IBDF and SUDAM
Government responsibilities for forestry in Amazonia have been divided between IBDF, EMBRAPA, SUDAM and other agencies. As mentioned above, IBDF no longer exists. Here we briefly describe the situation in 1988.

The responsibilities of IBDF included implementing and enforcing government policy and laws relating to logging and regeneration on private land, as well as fauna conservation and management of Flo. Na. Tapajós. IBDF Pará had staff in Belém, Santarém, Tapajós and elsewhere. However, IBDF had so few staff in Amazonia that they were quite unable, and presumably hardly expected, to supervise logging or enforce the laws. EMBRAPA had responsibility for inventories, for most research and for developing land-use systems, while SUDAM was responsible for implementing plans and projects. There was not always close collaboration between these agencies; presumably the new arrangement is designed to improve this situation. In general, the interest in TMF management and silviculture among Brazilian foresters has grown steadily in recent years, and several foresters now have significant practical field experience. However, the active interest is still mainly confined to the Amazonian research centres, without major backing from Brasilia. Much of the work started in the late 1970s and early 1980s has suffered from cuts in funding, and new projects have difficulty in getting started even when officially approved. Still, the technical basis is now much stronger than in the mid-1970s.

Bolivia

None of the reserved forests is subject to operational management for sustained timber production, but many individual management activities are being, or have been, implemented in different areas. Inventories and simple work plans have been made in many concessions, and the principles are well known. A forest management plan for Chimanes BP is being prepared, with areas allocated for several different land-use objectives according to their potential. The concept of forest reserves is well established in law, and the practical problems are well known.

The forest law requires all concession companies to carry out timber inventories before logging. Most companies have inventory data for their concessions. Inventories are usually carried out by forestry consultants, and are subject to approval by the company and CDF, but are reportedly of very variable levels of reliability. In some of the best-managed concessions, the inventory results are used for planning the harvesting; elsewhere, they are carried out more to fulfil the terms of the concession agreement than to aid logging and management.

Reconnaissance-level inventories have been carried out in Los Chimanes, in parts of Pando, Chiquitos, Velasco and Beni, and in the south of the country. In the late 1970s, a national forest mapping project using ERTS[8] imagery showed 50.8 per cent of the country to have forest cover. A new study of the functional classification of land and soils is now being undertaken, initially in Beni.

CDF is responsible for preparing management plans for the permanent production forests and is at present collaborating in the preparation of a plan for Los Chimanes. Individual concession holders are responsible for preparing plans before they are allocated full concessions and harvesting contracts. These plans should include an inventory and feasibility study, and should cover the harvesting, regenerating and conservation of the forest. Almost all the concession holders have carried out inventories in their areas, many have simple working plans or Planes de Ordenacion, and some (possibly fifteen) have completed the requirements, and have full, if short, management plans prepared along traditional lines. These companies have full concession contracts, but it is reported that none fully implements its own plans; in particular the diameter, area and volume limits are not generally respected. Most companies operate on annual or short-term licences. Thus, they have less long-term security, but also avoid legal responsibilities for management and protection.

Several management initiatives are now being started or planned. One of the most important of these is at Los Chimanes. Initial observations by CDF and the German Forestry Mission in the 1970s suggested that there was potential for management. Later, the World Bank, FAO and other agencies collaborated in preparatory studies towards establishing an integrated harvesting and management operation there. The FAO-CDF inventory formed part of this programme, and the area was subsequently declared a Bosque Permanente de Producción. Seven logging concessions have been issued. The Estación Biológica del Beni (EBB) was established in 1982 (135,000 ha) and it was declared a Biosphere Reserve in 1986. The objective is mainly research, conservation and the provision of a resource to maintain the Chimanes people. The forest and the station are now the subject of an international agreement for debt-exchange and integrated management. There are other proposals at Cochabamba (management plan and project), Choré (a project proposal for forest management to be located at Las Pavas Research Forest with emphasis on research and practical training), and Tumupasa (mixed land use including forest management).

Meanwhile, the reality in most of the forest areas is a gradual extension of logging activities, and of settlement, clearance and cultivation, all, since the 1960s, promoted by government and funded by banks. Over this the forestry authorities can have little or no control. Within MACA, there are plans for a reorganization to promote joint land-use planning.

The regional offices of CDF (or to an increasing extent the decentralized UTD-CDF CORDES offices) are responsible for logging licences, concessions and other forestry activities but they are not yet involved in natural-forest management.

In Santa Cruz, the main logging and timber companies fulfil their reforestation obligations by funding a separate plantation programme, the Proyecto Plantaciones Forestales (PPF) of the Camara Nacional Forestal (CNF), a private enterprise association. PPF has the largest operational plantation programme in the country and the plantation products have a ready market; they have also achieved the important result of showing that commercial plantations can be financially productive. However, the plantations will do little to contribute to future timber supplies because *Eucalyptus* is the main tree planted, and does nothing for natural forest regeneration.

Peru

Peru has one of the most comprehensive bodies of forestry laws, including the compulsory preparation of management plans. In addition, a substantial amount of practical (but so far largely unco-ordinated) experience is being built up in line/enrichment planting, and in natural regeneration (with and without interventions), combined with botanical and ecological studies. So far, these studies are almost entirely incorporated in projects of limited duration and not in operational management. Management plans are prepared by companies, and a major management-planning effort was made in BN von Humboldt with FAO assistance, but there are still no forests subject to operational management for sustained-yield timber production, nor any where annual timber extraction is based on rates of growth in the logged forests. In fact there is little information about the behaviour or productivity of the logged or secondary forests, and little experience of management decisions based on sustained-yield criteria, or of supervising and enforcing concession-contract obligations. Nevertheless, there are several concessions which manage their harvesting operations efficiently, and are concerned to ensure their future supplies. Genuine long-term forest management will shortly be feasible in some areas.

Many timber inventories have been carried out, in most of the areas covered by logging contracts, especially in those of more than 1,000 ha. The data are of very variable reliability, and are not necessarily used as the basis for organizing logging activities. Indeed very few detailed management inventories have been carried out. The most important was the inventory of 200,000 ha in BN von Humboldt as part of the UNDP/FAO Demonstration and Management Project.

Since 1975, the requirements for obtaining logging contracts for areas larger than 1,000 ha have included a management plan. The forestry regulations give full terms of reference for the plans.

Companies working in areas of less than 1,000 ha usually have a simple logging contract (Contrato de Extracción) with few legal responsibilities. The larger companies' responsibilities for management and regeneration are well defined in law, but in practice supervision and control have been very limited and implementation has been minimal. Companies logging in the Bosques de Libre Disponibilidad (BLD) are required to prepare feasibility studies and a harvesting plan. They are not obliged to carry out their own regeneration work and may pass this reponsibility to DGFF on payment of the Canon de Reforestación. Companies logging in Bosques Nacionales are further required to prepare a more detailed management plan. In fact, about twenty of these management plans have been written, mainly by companies operating in areas of more than 20,000 ha. The plans generally use as a model the plan prepared by FAO for BN von Humboldt in 1979 (but not implemented). They include division of the area into compartments (cuarteles), and a nominal forty-year cutting cycle.

In theory, DGFF should carry out a detailed inspection of each contract area at least once every two years, to ensure that the plans are being implemented. In practice, regular inspections have not yet been possible, but have started in some contract areas. Some timber companies, concerned about future timber supplies for their industrial investments, are now starting to base their decisions on the principle of sustained supplies.

Most of the silviculture and management work in TMF is on a research, demonstration or pilot scale. In spite of the existence of management plans with carefully programmed exploitation of existing resources, and forecasts of long-term yields, logging activities are, in practice, governed more by short-term market demands and access problems than by management principles. In Peru (as elsewhere), the low volume of timber extracted from the forest per unit area is often given as a reason why it is not economic to apply management practices.

In 1980, legal obligations for regeneration of logged forest were defined in the Canon de Reforestación. The work is carried out by Comités de Reforestación, mainly in the BLD, funded by a tax called the Costo de Reposición Forestal. It is reported that a total of 1,026 ha were planted between 1982 and 1986. Most of the planting is in strips (often 20 metres apart and 3–5 metres wide). Unwanted trees are felled and ring-barked and trees are planted at 5-metre intervals. Species planted include *Swietenia* and *Cedrela*. No information is readily available about survival or production in these planted areas. In any case, the annual areas regenerated by these methods are small in relation to the areas of annual harvesting and natural regeneration.

Ecuador

There are no examples of operational sustained-yield management, and many potential management techniques (such as silvicultural treatments, PSPs) have not yet been tried even on a research scale. However, as elsewhere, there are many isolated examples of management-related activities: many timber inventories; the management plan and work programme in Pichincha (oriented towards protection); the former system of logging concessions (now dormant); and the ongoing programme of selection and declaration of reserved areas, of which a few receive at least some modest protection and management.

The programmes of forest reservation for various objectives, and the agro-forestry activities now operating well at pilot-scale, are the best developed of any in these ITTO countries.

It is reported that a total of 6.85 million ha of TMF have had timber inventories at various scales. These inventories have served mainly to indicate the locations of the richest resources of timber, not as a basis for yield control or sustained-yield management, or even for determining fees or regulating the issue of licences or concessions.

There are no forest management plans designed for the sustained

production of timber in TMF (but some for protection and conservation areas). In law, the privately owned protective forests should
each have a management plan. Some of them are said to have simple
plans oriented to protection from fire, cutting and encroachment.
Previously, logging companies with forest concessions were required
to have management plans approved by the forestry service, and
many of them did so, although the provisions were seldom correctly
observed. All Patrimonio Forestal areas are expected to have
management plans in due course. However, the current emphasis is
on identifying and declaring new Patrimonio areas rather than on
protecting and managing those already declared.

Silviculture and management are the responsibility of the Department of Forest Management of DINAF. Although at present there
is no integrated, production-oriented management of reserved
forests, there have been some important developments in recent
years affecting future timber production, especially in forest
reservation, concessions and agroforestry.

In principle the areas of the Patrimonio Forestal may now be
allocated to whatever objective or management system is considered
most appropriate. The final report for Napo and Esmeraldas includes
suggested future objectives and land use for each individual block,
including nature and soil conservation, crop or livestock production, agroforestry and timber plantations; but the management of
the natural forest for the sustained production of timber was not
suggested for any areas. Unless staff and funds for protection and
management are made available, it is expected that the gradual
occupation, logging, clearance and cultivation of accessible forest
areas will continue even in Patrimonio areas. If, in future, DINAF
attempts to introduce sustainable natural-forest management for
timber production, it will presumably be in areas of Patrimonio
Forestal. At present, the options are open for using these areas for
other objectives, and for producing timber by other means.

Until the 1970s there were several licensed logging concessions
operating in defined areas with defined annual yields. The legal
and practical provisions were similar to those operating in many
other countries, and there were serious difficulties of control and
protection. Finally, government decided to cancel almost all existing
concessions. The principal reasons given were non-compliance with
the terms of licences, and the practical and political impossibility
of protecting concession areas from occupation and cultivation by
colonists.

It is not clear whether the situation was worse than in most other countries of Latin America which operate with concessions. The important point is that Ecuador now uses a system of short-term logging licences in unreserved forest. This has affected the quality and efficiency of logging operations, and has encouraged foresters to consider other ways of ensuring future long-term supplies of timber, particularly agroforestry.

The Fundación Natura, an autonomous conservation organization, is also concerned with natural forests and their conservation and management. The Bosque Protector Pasachoa is managed directly by Fundación Natura, with biological studies, reforestation and facilities for the public. The Fundación is also actively involved in aspects of national park management, and is collaborating with the Worldwide Fund for Nature (WWF) and ITTO in the co-ordination of a tropical forest information network in tropical America.

Protective forests have very low levels of management. These are areas where soil and water conservation are important in the regional or national interest and government is, therefore, disposed to finance and/or organize planting and other protective measures, even on privately owned land. Many landowners have arranged to have parts of their property declared protective forests, as this makes them eligible for subsidies for tree-planting and other investments, and for official assistance in protecting these unused parts of their land from occupation by peasants seeking territory for cultivation.

Agroforestry project activities have in general operated successfully and evolved steadily. Despite problems and setbacks, substantial progress has been made by staff and farmers, not only in developing techniques, but also in attracting the active interest and participation of the colonist-farmers. At present the work is tied to projects, with foreign grants or loans and technical assistance, and suffers from some typical limitations such as the uncertain future after a project ends.

Honduras

The great majority of forestry practices in Honduras have been oriented to the pine forests, but attention is now being paid to the broadleaved forests also. There is a growing recognition that their management and productivity will be increasingly important in the future.

Most inventories have been concerned mainly with the resources of pine timber, but substantial areas of TMF have also been subject to inventory. Since 1981, the INFONAC national forest inventory has gradually covered the country. TIMSA has also carried out inventories in the areas allocated to them for sustained-yield management.

The only important management plan in operation is that written for Olancho in 1977, concerned mainly with the pines; but a management plan for TIMSA, dealing with natural forest outside their existing concession was due to be implemented in mid-1988. This area is known as El Zapote, and contains 5,336 ha of TMF. The plan aims at a sustained supply of hardwood logs for the TIMSA timber and plywood mill. When implemented, this will be the first example of professional TMF management for sustained timber production.

No silvicultural systems or treatments, and no sustained-yield management and harvesting, have yet been applied in TMF. There are, at least locally, some administrative controls over logging, based on sustained-yield policies. The management of the existing TIMSA concession will be an attempt to carry out a full range of management activities. However, in recent years, TIMSA has not been harvesting in its concession and buys its timber from dealers outside its concession area.

The most persistent management-related activities have consisted of the work of projects carried out by COHDEFOR and the Canadian International Development Authority (CIDA), running initially from 1974 to 1984, and again from 1988. Project activities included reconnaissance and management inventories, feasibility studies for management and industrial development, boundary demarcation, industrial credit and institutional support, and, from 1978, protection and demarcation of existing forests. This work was combined with publicity and discussions with neighbouring communities. However, no legislation was passed to give formal legal status to these forests. The work ceased from about 1985 but will be resumed from 1988.

The UK Overseas Development Authority (ODA) has also provided technical assistance to promote the rational management of TMF from 1983 up to the present. During this time, substantial advances have been made in forest laws and policy, and in new strategies for managing the broadleaved forests.

Since 1983 COHDEFOR has been promoting Areas de Manejo Integrado (AMI), with FAO assistance. These consist of a kind of social forestry system, involving the planned and rational use of the forest by local communities using simple techniques, including timber production by pit-sawyers. There are now about fifty AMI, each of between 1,000 and 10,000 ha, and the number is growing.

Another management model is the Sistema Industrial Forestal Energético Social (SIFES), for complete utilization of the forest resource to supply energy as well as timber. One such project, near Comayagua, has a small sawmill powered by a steam generator using pine waste wood. The system may be extended to TMF. From 1988 new projects, with assistance from Canada and the US, will include forest management components.

New management-oriented logging rules have recently been prepared by ODA/COHDEFOR (1988) for harvesting in broadleaved forests. These are designed to increase the intensity of harvesting and the number of species logged and used. The rules require that all commercial logs of compulsory species must be cut, with detailed definitions. Logging contracts in future are expected to stipulate that 3 cu. metres of lesser-used species must be harvested for each cubic metre of high-value species.

COHDEFOR has substantial responsibilities for planning and controlling the harvesting of timber from private forests, and for management in these areas. In practice, COHDEFOR is able to promote a certain amount of management in some areas of private pine forest, but not in TMF.

In effect, the accessible areas of TMF are steadily being converted to cultivated land or pasture in an unplanned piecemeal fashion by new farming families, by farmers who have lost their jobs or land, by refugee-immigrants and by landowners and cattle-owners who contract workers to clear and burn forests and also buy plots from farmers. Often, the clearing is preceded by a selective harvesting of the most valued timber. Thus substantial areas of forest are cleared and incorporated into the larger landholdings. On the other hand, during the past twenty-five years, a process of land reform has resulted in many under-used estates being allocated to small farmers or co-operatives.

RESEARCH

The largest amount of published information from research studies on TMF relates to botanical composition and ecological processes. In addition, there are important silvicultural treatment experiments in Trinidad (Arena forest, 1956) and in Brazil (Curuá-Una from 1959; Tapajós; Jarí; INPA, Manaus), but none of these is directly tied to existing, operational silvicultural programmes. Analysis of the Brazilian data is under way and is expected to increase understanding of growth processes and reactions to interventions, but so far the various studies are little co-ordinated and little related to the realities of management. Plantation and enrichment trials have been carried out in many areas. Some conspicuously successful results have been achieved in Tapajós, Brazil, and in von Humboldt, Peru. Permanent sample plots for management purposes have been established in Trinidad. Permanent plots for research and treatment trials have been established in many areas of Brazil and at least three plots exist in unlogged TMF in Honduras. In Peru, many TMF plots were established in von Humboldt during the FAO project, but it is reported that none survives.

Table 4.8: Volume of logs and number of logging contracts

	Annual volume (cu. metres)	No. of logging contracts
Trinidad	50,000	800
Brazil	4,000,000	?
Bolivia	115,000	195
Peru	?	3,000
Ecuador	450,000	4,000

LOGGING

Official estimates of industrial log production from TMF areas are shown in Table 4.8. The actual volume of logs entering industries is certainly larger than the official figures, perhaps double. Logging is highly selective, to the extent that few species, notably *Swietenia macrophylla*, are consistently taken. However, most industries use about five to twenty species. Average logging intensities are usually 2–5 cu. metres per hectare, but often as low as 1 cu. metre. Locally they exceed 25 cu. metres in certain well-organized long-term concessions.

Fees charged by governments and forestry services vary very widely, as shown in Table 4.9. This extreme variation partly reflects bureaucratic delays in revising laws and regulations but also, undoubtedly, differences in attitudes of national governments towards the exploitation of natural resources and the partition of benefits. Certainly, the higher fees charged in Honduras are partly the result of the self-financing, para-statal status of the forestry service COHDEFOR, and are related to current initiatives to improve forest management there. These charges refer to those licensed operations which pay fees to government, although their cutting situations vary from exclusive concessions to private or settled land. The issuing of logging licences or concessions involves a wide range of systems: in Bolivia, 195 concessions cover over 22 million ha (out of a total of over 40 million ha of TMF); in Ecuador, about 4,000 cutting licences are issued annually, for a much smaller total forest area (perhaps 12 million ha); even more extreme, in Trinidad, there are about 800 licensed loggers for a productive TMF estate of about 75,000 ha.

Table 4.9: Fees per cubic metre

	US$
Trinidad	3.00-4.50
Brazil	0.55 (regeneration fee)
Bolivia	5.00 (+2.5 regeneration fee)
Peru	0.01-0.20
Ecuador	0.40
Honduras	7.50-40.00

These experiences demonstrate that a system of abundant small loggers may function adequately, as in Trinidad, if well controlled, and that any system is defective if poorly controlled.

INDUSTRIES

Much less information could be collected on timber-using industries than on other elements of the study. In general, the level of technology and efficiency of the industries is related to the same factors which affect the efficiency of logging and indeed of management: the abundance of the resource and the distance and size of the markets. All member countries have a large majority of relatively small sawmills, wasteful of raw materials and manpower (though doubtless profitable). There are apparently relatively few modernized, technically efficient sawmills such as are found locally in South East Asia.

Official data show the following numbers of log-using industries in member countries.

Table 4.10: Number of log-using industries

	Sawmills	Plywood and veneer mills
Trinidad	64	–
Brazil (Amazon)	2,000 (approx.)	100 (approx.)
Bolivia	219	8
Peru	>500	11
Ecuador	456	12
Honduras (using TMF species)	5	2

TRAINING

All member countries (except Trinidad, which sends its students abroad) have university facilities for training forestry graduates and often also technicians. Admissions are generous (in terms of numbers) in relation to actual job prospects. Thus, the numbers being trained are often larger than are needed to maintain the current half-hearted controls over logging in the absence of serious attempts at forest management. The number of trained foresters, however, is far below that required to manage the forests effectively. Most of the universities do not include management studies for the preparation of forest or land-management plans.

On the other hand, the facilities for training of forest workers, guards and *peritos* are generally entirely inadequate for either actual or potential employment. Forest workers usually receive no special training. In some countries (e.g. Ecuador) the number of trained graduates qualifying annually is larger than the number of forest guards.

There are few unfilled posts, although there are widespread expressions of the "need" for more staff at all levels. This is because government funding for most of the planned, proposed and "needed" forestry activities (such as management and protection) is often skeletal. It is often very limited even for those forestry activities which are actually carried out, such as the establishment of plantations and control of harvesting. Employment prospects in forestry will remain limited until existing activities are carried out more intensively and other elements of management and extension are taken seriously. When that time comes, most regions will experience a shortage of staff with adequate training and, especially, experience in management.

The relatively generous provision of university training, in comparison to training for guards and workers, arises partly from the different processes of funding and ways of reaching decisions. Often, the universities have a high degree of autonomy in their power to spend their annual incomes. The decision to found a new faculty of forestry may be internal; or may depend upon successfully convincing the ministry or a funding agency of the benefits to be derived from training more graduates, or of the "need" for more graduates. This is often done without a detailed study of the present and future job market and certainly without financial or professional responsibility for ensuring suitable employment for the graduates.

On the other hand, technical training schools are more often funded or staffed directly by government, and are usually managed within the forestry services, or by the Ministry of Education in close collaboration with the forestry services, with the expectation that most of the qualified technical staff will be employed in those services. Under these circumstances, training is often more directly related, both in curriculum and numbers, to the needs and opportunities for employment.

NOTES

1. This chapter is condensed from the report which IIED presented to ITTO, in which a separate account was given of each member country. More detailed information can be obtained from that report.

2. The names of the many organizations in the region, and their abbreviations, are given in an Appendix to this chapter.

3. Eden, J., "Silvicultural and agroforestry developments in the Amazon basin of Brazil", *Commonwealth Forestry Review*, 61, 3 (1982), estimated the total Amazonian TMF at 260 million ha including seasonally flooded forest, while FAO, *Los recorsos forestales de la America tropical* (Rome: FAO, 1981, UN 32/6.1301-78-04 Inf. Tec. 1), gave a total of 340 million ha.

4. FAO, op. cit.
5. FAO, *Intensive Multiple-use Forest Management in the Tropics*, FAO Forestry Paper No. 55 (Rome: FAO, 1985).
6. Pitt, J. *Application of Silvicultural Methods to the Forest of the Amazon*, FAO Report No. 1337 (Rome: FAO, 1961); Pitt, J. *Relatoria ao Governo do Brasil sobre aplicacão de métodos silviculturais a algunas florestais da Amazónia* (Belém: SUDAM, 1969).
7. Wood, T.W.W., *Plano de manejo y desarollo forestal do Tapajós* (Brasilia: IBDF, 1980); also FO:BRA/78/003 Technical Report No. 2(e).
8. Etudes des Ressources Terrestres pour Satellite.

APPENDIX: ABBREVIATIONS

Trinidad and Tobago

FD	Forestry Department
FR	Forest Reserve
FRIM	Forest Research, Investigation and Management section
PSP	Permanent Sample Plot

Brazil

CENARGEN	Centro Nacional de Recursos Geneticos, EMBRAPA, Brasilia
CNPq	Conselho Nacional de Desenvolvimento Cientifico e Tecnologico
CPATU	Centro de Pesquisa Agropecuaria do Tropico Umido, Belém, PA
CTM	Centro de Tecnologia de Madera, Santarém
EMBRAPA	Empresa Brasileira de Pesquisa Agropecuaria
FCAP	Faculdade de Ciências Agropecuarias do Pará, Belém
FUNAI	Fundaçaõ Nacional do Indio
FUNTAC	Fundação de Tecnologia do Estado do Acre, Rio Branco
IBDF	Instituto Brasileiro de Desenvolvimento Florestal
INPA	Instituto Nacional de Pesquisas da Amazonia, Manaus

PRODEPEF	Projeto de Desenvolvimento e Pesquisa Florestal
SUDAM	Superintendencia do Desenvolvimento da Amazonia

Bolivia

BP	Bosques Permanentes de Producción
CDF	Centro de Desarrollo Forestal
CNF	Camara Nacional Forestal
CONRA	Consejo Nacional de Reforma Agraria
CORDES	Corporación Desarrollo
EBB	Estación Biológica del Beni
INC	Instituto Nacional de Colonización
MACA	Ministerio de Asuntos Campesinos y Agropecuarios
PPF	Proyecto Plantaciones Forestales
RFI	Reservas Forestales de Inmovilización
UTD	Unidad Técnica Desconcentrala

Peru

APODESA	Apoyo a la política de desarrollo en la Selva Alta, INADE
BLD	Bosques de Libre Disponibilidad
BN	Bosques Nacionales
BP	Bosques de Protección
CENFOR	Centros Forestales
CEPID	Centro de Estudios de Proyectos de Inversión y Desarrollo, UNA
CONYTEC	Consejo Nacional de Ciencia y Tecnología
CORDES	Corporación Desarrollo
COTESU	Cooperación Técnica Suiza
DGFF	Dirección General Forestal y de Fauna
IAP	Instituto de Investigaciones de la Amazonia Peruana
INADE	Instituto Nacional de Desarrollo, Ministerio de la Presidencia
INFOR	Instituto Nacional Forestal y de Fauna
INIAA	Instituto Nacional de Investigaciones Agropecuarias y Agroindustrias
JICA	Japanese International Co-operation Agency

PEPP	Proyecto Especial Pichis-Palcazu, INADE
UC	Unidades de Conservación
UNA	Universidad Nacional Agraria La Molina, Lima

Ecuador

AIMA	Asociación de Industrias Madereras del Ecuador
BNF	Banco Nacional de Fomento
CLIRSEN	Centro de Levantimientos Integrados de Recursos Naturales por Sensores Remotos
CORMADERA	Corporación de Desarrollo del Sector Forestal y Maderero
DEFORNO	Desarrollo Forestal del Noroccidente
DINAF	Dirección Nacional Forestal
EMDEFOR	Empresa de Desarrollo Forestal
FONAFOR	Fondo Nacional de Forestación y Reforestación
IERAC	Instituto Ecuatoriano de Reforma Agraria y Colonización
INERH	Instituto Ecuatoriano de Recursos Hidráulicos
INIAP	Instituto Nacional de Investigaciones Agropecuarias
INICEL	Instituto Ecuatoriano de Electricidad
MAG	Ministerio de Agricultura y Ganadería
PRONAREG	Programa Nacional de Regionalización Agraria

Honduras

AHE	Asociación Hondureña de Ecología
AMI	Areas de Manejo Integrado
CENIFA	Centro Nacional de Investigación Forestal Aplicada
COATLAHL	Cooperativa Agroforestal Colón, Atlantida, Honduras
COHDEFOR	Corporación Hondureña de Desarrollo Forestal
CORFINO	Corporación Forestal Industrial de Olancho
CURLA	Centro Universitario Regional del Litoral Atlántico
DIGERENARE	Dirección General de Recursos Naturales Renovables

DISA	Danlí Industrial, SA
ESNACIFOR	Escuela Nacional de Ciencias Forestales
INFONAC	Inventario Forestal Nacional
INIA	Instituto Nacional Agrário
SIFES	Sistema Industrial Forestal Energéticos Sociales
TIMSA	Tela Industrial de Madera, SA

5. Asia
P.F. Burgess

FOREST POLICY

The Forests and Forest Protection[1]

India, West Malaysia, Sarawak, Sabah and Indonesia have a policy of reservation of permanent forests for both productive and protective purposes. Except in Indonesia this is by a two-part system of gazette notification with a public judicial inquiry between the two. In West Malaysia, Sarawak and Sabah, dereservation of forest reserves, almost unheard-of before the early 1970s, had become commonplace by the early 1980s largely to accommodate expansion of permanent agriculture in the lowlands. In Sabah some dereservation also took place ostensibly for agricultural development which then often did not materialize, so that the forest was logged over, fortunes made, and the land abandoned.

For a period after independence the dereservation of forest reserves was relatively easy in some states of Malaysia (land and forests are state, not federal, matters), the agreement of the Chief Minister without the backing of the Executive Council being sufficient. This has recently been corrected and the agreement of a senior minister in council is now once again required. In Peninsular Malaysia it has also recently been declared to be government policy not to alienate forest land until substantial progress has been made with developing land already alienated, and to replace forest excised from reserves with new areas of state-owned forest where this is available.

In Indonesia the forests were classified by the Department of Forestry in 1983 into:

(a) permanent forest land
    protection and conservation forests;
    limited production forests (forests where there are constraints to full commercial use);
    production forests;

(b) conversion forests (i.e. forests for eventual alienation).

A process of land-use classification by consensus was then developed at provincial level where representatives of the interests involved met to pool available information on the forest lands, to compare the requirements of each sector for land for development and to agree jointly on the land use for each area. It is not clear to what extent the land-use classification is based on accurate survey information and may therefore be subsequently modified, but it hardly seems likely that the system will ensure for the permanent forests the long-term protection from alienation which is essential for sustained-yield management.[2]

In Thailand and the Philippines all land not alienated is declared to be forest land. Thus government takes no positive action to constitute forest reserves, the forests being merely the residual state or crown land. In some cases it may not even be forested. As a result forests are not regarded as sacrosanct, they may be alienated whenever a potentially more valuable use develops, and clearing of forests for shifting cultivation tends to be regarded lightly.

In Papua-New Guinea almost all land (97 per cent) is owned by either clans or individuals who for centuries may have made no use of it except for hunting or subsistence farming. The forest utilization rights may be bought out by government for limited periods with the agreement of the owners, but the rights are almost never ceded for a second cut. There are thus no permanent forests in Papua-New Guinea except where government has bought both timber and land rights; this is only possible over limited areas, for example for experimental plantations.

Control of the rate of forest utilization

In Peninsular Malaysia, Sarawak, Sabah, the Philippines and Indonesia the national forest policy provides for the sustained-yield management of the permanent forest estate, and for the supply of forest produce for agricultural, domestic and industrial purposes within the country and for export. There is, however, no definition of what quantity or quality of forest produce shall be so provided, nor is there any indication of the volume of exports which should be sustained after domestic requirements have been met. Consequently, as development of alienable and disposable land for agriculture proceeds, and because timber production from the permanent forest estate exceeds domestic requirements, there are temporary floods of exports which eventually decline as forests are worked out. By this time, however, an influential industry has developed with huge investment in roads, extraction plant, mills and infrastructure, and there is immense pressure on government, which is generally the sole forest owner, to permit extraction to continue in the permanent forests at an unsustainable rate.

As a result, total timber production in Peninsular Malaysia will within a short period be insufficient to supply even the internal requirements of the country. Thailand, long a major exporter of teak and yang, is now importing logs from Burma and, paradoxically, from Peninsular Malaysia. Moreover, the standard of timber available within the more developed parts of the region has declined sharply during the last decade both in species and grade; high-quality dipterocarp timber from virgin stands will soon be a thing of the past in all but a very small part of the region.

In Thailand there is no clear commitment to sustained-yield management in the national forest policy and indeed it has been stated that the national forest administration shall be reorganized in line with the changing quality and quantity of forest resources and the environment. Timber extraction from state forests was suspended in late 1988.

In Papua-New Guinea no true sustained-yield management can be practised by government as timber rights can only be acquired for a maximum of forty years; indeed replanting and even natural regeneration are resisted by landowners, since these may establish a right to transfer of ownership of the land.

THE FORESTS

Classification by area

The extent of forest of different types is summarized in Table 5.1.

With some minor exceptions, such as peat swamp forests in Peninsular Malaysia and Sabah and upland pine forests in the Philippines, the production forests of Peninsular Malaysia, Sarawak, Sabah, Indonesia and the Philippines are all dipterocarp forests with varying proportions of soft-wooded *Shorea* spp., hard-wooded *Shorea* spp., *Parashorea* spp., *Dipterocarpus* spp., *Dryobalanops* spp. and other minor genera of the Dipterocarpaceae.

Table 5.1: Forest areas in ITTO member countries in Asia (sq. km)

	Total land area	Total forest area	Permanent protection forest	Permanent production forest	Conversion production forest	Production forest (all)	Virgin forest remaining
Thailand	513,115	142,958	?	?	?	?	?
Malaysia							
Peninsular	131,596	63,532	10,679	34,507	9,100	43,607	9,600
Sabah	73,711	44,869	6,000	29,984	4,080	34,064	7,815
Sarawak	123,253	94,384	24,200	32,400	37,784	70,184	50,387
Philippines	300,000	63,830	16,800	44,030	0	44,030	10,420
Papua-New Guinea	468,860	359,900	?	?	?	?	?
Indonesia	1,930,270	1,439,700	490,410	338,666	305,370	644,036	524,000
Total	**3,540,805**	**2,209,173**	**548,089**	**479,587**	**356,334**	**835,921**	**602,222**

These forests also contain timber-producing species of the families Leguminosae, Burseraceae, Meliaceae, Anacardiaceae, Apocynaceae, Datiscaceae, Sapotaceae, Sterculiaceae and many others. The conifer *Agathis* also occurs locally. In Sarawak and parts of Indonesia there are extensive peat swamp forests producing two valuable non-dipterocarp timbers, *Gonystylus bancanus* (ramin) and *Dactylocladus stenostachys* (jongkong). In parts of Indonesia, for example on Halmahera, there are forests very rich in *Agathis*. To the east of Wallace's Line the dipterocarps are generally poorly represented, though *Anisoptera* and *Hopea* spp. occur in Papua-New Guinea. The forest there is much more mixed and, consequently, of less commercial value, though exceedingly rich forests such as the *Terminalia brassii* swamps of Bougainville may occur locally. In Thailand the moist tropical forests grade to the north and east into drier dipterocarp and teak forests.

Selection, Demarcation and Protection

The selection, demarcation and protection of permanent production and protection forests varies widely within the region, the difference often being inherited from previous administrations. There is a common system shared by Peninsular Malaysia, Sarawak and Sabah, since senior staff in Sabah and Sarawak at the start of the active period of forest reservation came largely from the Malayan Forest Service.

In Peninsular Malaysia a systematic programme of exploration, selection, survey and demarcation of both production and protection forest reserves has been followed since the early 1900s, though many of the lowland reserves were lost to agriculture through dereservation in the years immediately following independence. The situation has now been largely stabilized because greater emphasis is being placed on the proper development of state land and alienated land before any further dereservation is considered. Moreover, the stimulus for land alienation has been much reduced now that there is a real prospect of rubber, palm oil and cocoa being overproduced on a world scale. In addition, the principle has been accepted that excisions should be balanced by the reservation of replacement forest. As a result, though forestry has in general been pushed into more mountainous and remote areas, where both exploitation and regeneration are more difficult, this is, at least partly, balanced by increased security of tenure.

In Sarawak forest reservation has been complicated by the existence of native customary rights over large areas of forest. This has led to the creation of a class of "protected forest" – a kind of halfway stage to forest reservation – where, pending detailed land-use surveys, wide rights are conceded to take forest produce for domestic use and to pasture cattle. Forest reservation was far from complete at the time of independence and, of late, the Forest Department has had great difficulty in securing the co-operation of the administration in the expeditious conduct of the inquiries prescribed under the Forest Ordinance as part of the reservation procedure. There has also been organized resistance by hill tribes to the extension of logging into remote forest areas and this has inevitably made forest reservation unpopular in official circles.

This situation is reflected in the data given in Figure 5.2 for the reserved forests, excluding those in the peat swamps.

Table 5.2: Reserved forest in Sarawak

Forest reserve	4,762 sq. km
Protected reserve	34,064 .. ..
State land forest	48,437 .. ..

Clearly much remains to be done in the selection and complete reservation of forests.

In Sabah the Forest Department embarked during the late 1950s and early 1960s on an ambitious and comprehensive programme of forest reservation before land classification studies had been completed. The process was strengthened by a legal ruling that all long-term concession areas should be constituted as forest reserves in order to protect the rights of the concessionaires. But, inevitably, many of these early concessions were located in some of the most accessible forest. When overseas concessions expired and were not renewed in the 1970s, there was pressure to dereserve the more accessible parts from the permanent forest estate.

Table 5.3: Forest reserves in Sabah

Commercial (i.e. production) forest reserves	26,746 sq. km
Protective and domestic forests	3,576 .. ..
Total	**30,322 .. ..**

Reserves in Sabah in 1987 covered the areas shown in Table 5.3. The total of 30,322 sq. km compares favourably with 28,221 sq km under reservation in 1962; the dereserved areas have been more than replaced by further reservation, but in more inaccessible forests.

In Papua-New Guinea no system of forest reservation exists, nor can it under the existing land tenure system. In 1987 the situation was as shown in Table 5.4.

Thinking in Papua-New Guinea is now directed towards making the Forest Department into a technical extension service to private forest owners, comparable to that provided for agriculture. Clearly attempts to operate a department on conventional lines are unlikely to succeed where government does not own the forest.

Table 5.4: Forest reservation in Papua-New Guinea

Forested area	364,000 sq. km
of which operable	150,000 .. ..
logging rights acquired	37,000 .. ..
Government forests	3,211 .. ..

In Indonesia there is no formal reservation of forests but a land-use plan is developed for each province by a process of consensus between different interests. The situation in 1985/6 is presented in Table 5.5.

The dominant position of Indonesia as a holder of forest resources in the region must be emphasized. The figures in Tables 5.1 and 5.5 suggest that Indonesia has 65 per cent of the total forest area in the countries included in this survey; 80 per cent of the production forest; and 90 per cent of the virgin stands. The figures for Indonesia require verification, however, for there is, as yet, no national forest inventory. There is a long history of shifting cultivation in Sumatra, Kalimantan and Irian Jaya, and evidence of much natural destruction by drought, fire, volcanic activity and mass earth movement in Irian Jaya and Papua-New Guinea.

In Thailand and the Philippines, where forest land is, by definition, the balance of land not alienated, the selection of production forests has largely been determined by the requirements of timber concessionaires. This is not necessarily a bad thing, since it means that the first forests to be selected were often the most valuable and

Table 5.5: Forest areas in the provinces of Indonesia

Province	Total land area (sq. km)	Total forest area (sq. km)	Permanent forest area (sq. km)	% of total land area	Forest for alienation (sq. km)	Alienated land (sq. km)
Sumatra	469,493	302,072	251,757	53.6	50,315	167,421
Java	132,190	30,133	30,133	22.8	0	102,057
Kalimantan	548,247	449,677	366,743	66.9	82,934	98,570
Sulawesi	196,615	132,846	112,914	57.4	19,932	63,769
Bali	5,632	1,257	1,257	22.3	0	4,375
Nusa Tenggara	67,542	55,475	25,500	37.8	24,975	12,067
Maluku	85,728	55,333	50,969	59.4	4,364	30,395
East Timor	14,609	6,998	6,898	47.2	100	7,611
Irian Jaya	410,660	405,915	288,161	70.2	117,754	4,745
Total	**1,930,716**	**1,439,706**	**1,134,332**	**58.8**	**305,374**	**491,010**

were therefore afforded some protection by the concessionaires. In Thailand, however, the forests, once logged, have been so badly damaged by shifting cultivation and illegal logging, exacerbated by the withdrawal of the protection afforded by the concessionaires, that they have become scarcely worth further maintenance.

The same has happened in the Philippines. At present, forests are being surveyed and demarcated for permanent protection under the Land Classification and Survey Programme, run by the National Mapping and Resource Information Authority, on which foresters are represented. The programme is now said to have covered more than 90 per cent of the Philippines, but this estimate may be over-ambitious. Concessionaires are obliged to survey, demarcate and protect their forests; and, in theory at least, the permitted coupe is reduced if illegal clearing and logging occur.

National Forest Inventory

There is no national forest inventory in either Indonesia or Papua-New Guinea.

In the Philippines a forest inventory has been carried out covering all forest lands. Volumes of timber by region are published in *Philippine Forest Statistics, 1986*. This is being updated by a Filipino-German team.

In Thailand there is a Forest Inventory Section within the Royal Forest Department. Preliminary inventory studies started with the assistance of FAO in 1953 and continued until 1969. A full inventory by the Royal Forest Department was started in 1969, and by 1976 over three-quarters had been completed, the balance being in sensitive areas where work was not possible. Intensive inventory for working plans has continued in the northern provinces.

The first forest inventory on a national scale in Peninsular Malaysia was carried out from 1970 to 1972 by FAO. This inventory was updated by the second national forest inventory in 1981-2. Stand and stock tables were produced by species and species groups, diameter classes and quality classes, together with stock maps at scales of 1:250,000 and 1:750,000.

In Sarawak a forest resource inventory was carried out in 1969–72 over 1.2 million ha (about 13 per cent of the forest area). This work was done also by FAO. The Forest Department has continued this inventory and, by the end of 1985, 3.56 million ha had been covered, just over 40 per cent of the total forest area.

The first national forest inventory of Sabah was carried out in 1969–72 under a Canadian bilateral aid programme. The Forest Department has carried out an inventory of the logged-over forests from 1986 to 1988.

FOREST OPERATIONS AND MANAGEMENT

Concession agreements

In all the countries in the region forests are let out for logging under concession. They differ considerably, however, in the proportions of their forest that are under concession, in the period and general terms of the agreement and in methods of selecting concessionaires. Some countries, too, have a longer history of exploitation than others. Some of these matters are discussed in this section.

In Thailand the exploitation of the forests is largely in the hands of the Forest Industries Organization (FIO), a state body of which the president is the Minister for Forests, the chairman the Director General of the Royal Forest Department and the board members all senior government officials. The FIO also has a controlling interest in the Thai Plywood Company and a 46 per cent share in the Provincial Logging Companies (Changwad Co. Ltd).

Table 5.6: Areas under concession in Thailand

	Teak (sq. km)	Other forest (sq. km)
Forest Industries Organization	24,860	16,212
War Veterans Association	6,072	9,707
Provincial Logging Companies		126,202
Thai Plywood Co.		5,148
State railways		2,674
Miscellaneous private firms		3,313
Total	30,932	163,256

The areas under concession are summarized in Table 5.6. The figures are revealing. They show that the area under teak concession is roughly equal to the total area of teak forest (32,247 sq. km) but the area under concession in other forests exceeds the total area of those forests by some 50,000 sq. km, suggesting that the concessions contain considerable areas of non-forest land. The concession agreements started in 1968 and are for thirty years; so some two-thirds of the forest will now have been worked over – not of course for the first time! Concession agreements are renewable with the approval of government, but there is no absolute right of renewal.

In Peninsular Malaysia there are some seventy-three concession agreements in force in the permanent forests and a further twenty-two in conversion forests. These vary in period from three to sixty years, with the majority (sixty-five) for under fifteen years. The earliest concession agreements were issued in about 1965 with most of the remainder in the 1970s and early 1980s. The total area under concession agreement in forest reserves covers 10,607 sq. km, an area equivalent to just over 30 per cent of the area of permanent production forest. The combined annual coupe in these concessions amounts to 44,487 ha, representing a mean cutting cycle of 23.8 years.

Individual concession areas are not necessarily sustained-yield units, nor are the total permanent forest concession areas in each state worked on a sustained-yield basis. In fact the average cutting cycle varies from five years in the state of Negri Sembilan to thirty-eight years in Pahang, a variation due in part to some of the forest reserves being worked for alienation. Nevertheless, the coupes in Peninsular Malaysia as a whole are not being fully cut each year and there is an accumulated undercut of some 137,000 ha. Several large concessions in the states of Pahang and Trengganu have accumulated uncut coupe for many years because of inefficient working. If they could assemble an efficient operation and if market conditions were right, it would be possible for them to work the uncut coupe in a very short period of time.

In Sarawak 715,023 ha of peat swamp and 4,154,932 ha of upland forest were under long-term agreements at the end of 1986. In addition, 719,859 ha of peat swamp and 810,227 ha of upland forest were under annual licence. The number of long-term licences totalled 341 with a mean area of 14,280 ha.

In Sabah the areas under concession agreement were as shown in Table 5.7. Of these only the Sabah Foundation area is under sustained-yield management. The area under concession agreement amounts to 59 per cent of the area of production forest in the state. Many of the smaller concessions, though within the permanent forest estate, are operating on relatively short-term agreements and are licensed to complete logging within the period of the agreement.

Sabah Kuwait Timbers, for example, has 77,245 ha in the Sapulut Forest Reserve and is permitted to complete logging within eight years, after which the forest will be closed, temporarily at least.

Table 5.7: Areas under concession agreement in Sabah

Sabah Foundation 100-year agreement	1,070,866 ha
Sabah Forest Industries (alienated)	300,034 ..
Others	633,324 ..
Total	**2,004,224 ..**

In the Philippines most concessions are for twenty-five years but some for as little as five. As they expire they will be replaced by timber production sharing agreements (TPSA), in which government becomes the major partner and timber is sold at prevailing stumpage rates. The last of the existing concessions was issued in 1983/4. By the end of 1986 a total of 142 timber licence agreements were in force covering 5,675,358 ha with an annual coupe of either 8,231,483 cu. metres or 82,291 ha, whichever is achieved first. This assumes a yield of 100 cu. metres per hectare, a figure which seems high for an average, and a cutting cycle of seventy years. The mean area covered by an agreement was about 40,000 ha.

Table 5.8: Extraction licences in Papua-New Guinea

	Area in hectares	*Licences*
20-year agreements	353,366	4
10-year agreements	515,652	32
5-year agreements	81,754	19
Shorter periods	225,356	
Total	**1,176,128**	

In Papua-New Guinea licences in force at the end of 1985 were as shown in Table 5.8. All agreements require the negotiation of a timber rights sale by the landowners to government, and these agreements can only be concluded for a maximum period of forty years.

In Indonesia concessions have been issued for over 53,374,000 ha, there being 521 such licences averaging 102,445 ha each. Over 84 per cent of the area licensed is in Sumatra and Kalimantan. In order to restrict the opportunities for illegal logging, licences are not normally issued for less than 35,000 ha. The total area under licence is slightly larger than the total estimated area of operable forest; there thus appears to be little scope for further licensing, unless the "limited production forests" (i.e. those less accessible) are tapped. Most agreements are for a twenty-year period, normally renewable on expiry if work has been satisfactory. From 1989 much stricter control will be enforced over concession working in Indonesia and a substantial number of concessions have been cancelled for non-compliance with cutting rules.

It is of interest to summarize the area under concession agreement in the region and to compare it with the area of production forest (Table 5.9). It will be seen that the proportion of total forest under concession is just over 66 per cent. The most notable features in this survey of concessions are: the issue of concessions in Thailand over an area substantially greater than the total forest area; the huge area under concessions in Indonesia, over twice that for Malaysia, the Philippines and Papua-New Guinea combined; and the great apparent potential for the further issue of concessions in Papua-New Guinea.

Table 5.9: Areas of production forest and areas under concession in Asia (1986 figures)

Country	Area of production forest (ha)	Area under concession (ha)
Thailand	14,295,800	19,418,800
Malaysia		
Peninsular	4,360,700	1,060,700
Sarawak	7,018,400	6,400,041
Sabah	3,406,400	2,004,224
Philippines	4,403,000	5,675,358
Papua-New Guinea	35,990,000	1,176,128
Indonesia	64,403,600	53,374,000
Total	**133,877,900**	**89,109,251**

Long-term concessions are covered by legal agreements in all countries. Schmithüsen in FAO Forestry Paper No. 1 has detailed the forest management provisions which need to be included in a long-term concession agreement.[3] Essential elements are:

- a statement of the maximum allowable annual cut (preferably by both volume and area)
- a definition by map and boundary description of the concession area
- procedure for opening and closing annual coupes
- control by the forest owner over main and possibly spur road location and construction standards
- minimum girth limits
- marking procedure for fellings (this to include directional felling)
- slope limits
- list of prohibited, obligatory and optional species
- rules for conducting felling and extraction operations
- obligatory milling clause
- survey and demarcation of concession and coupe boundaries
- survey of roads and fellings and maintenance of progress maps
- scaling procedure
- fixed penalties for infringements
- concessionaires' obligations for protection of the forest against illegal clearing and logging
- prohibited extraction methods
- minimum standards for residual forest and definition of acceptable standards of logging damage
- concessionaires' obligations to carry out silvicultural work
- free access to all operations at any time by forest department staff
- employment of professional foresters by concessionaire
- mapping and forest inventory procedures for the concession as a whole and for coupes before and after exploitation.

With the exception of those in Papua-New Guinea, the concession agreements in the region are generally reasonably adequate or could be made so with relatively little modification.

With a view to increasing the independence of forest department staff and, thus, their ability to "get tough" with recalcitrant concessionaires, it is desirable that concession clauses obliging the operator to provide housing, transport and other facilities for them should either be deleted (perhaps in exchange for more onerous clauses to control logging) or not enforced, so that they are independent of the logging operation.

While, in general, concession agreements have adequate provision to ensure the orderly and efficient sustainable management of forests, the reality is less satisfactory, due to weak enforcement by many forest departments. The reasons for this are complex and lie at the heart of the whole problem of lack of sustainable management; this matter will be examined further in the discussion at the end of this chapter.

Concessions have commonly been issued for periods of 21 to 25 years, although the minimum realistic felling cycle is 30 years and the rotation 60 years. This means that concessionaires have no interest in the forest once they have logged it and will not, therefore, take very great pains to log it carefully; nor will they bother much to protect worked-over coupes from illegal felling or clearing, or protect roads from erosion damage. They will have no interest in planting up blanks, since they will have no claim to harvest that which they have planted.

The renewal of concessions is a very chancy business, and may depend on such factors as the election of a particular political party; no operator can rely on it. In the Philippines the Forest Department undertakes an audit of the performance of each concessionaire every three years and the score goes on record to be used when renewal is being considered. In Indonesia concessions are said generally to be renewed if work has been satisfactory. But it is unrealistic to suppose that the Bureau of Forestry in the Philippines would succeed in denying the renewal of a concession if it were politically expedient that it should be renewed. Nor could the bureau effectively promote the renewal of a concession if there were political reasons against it. The days are gone when forest officers had control over the selection of forest operators. Concessions are valuable assets and in all cases encountered in this study they are disposed of by politicians, generally without the advice of foresters.

In Sabah the only long-term concession is now in the process of being renewed for 100 years. The lessee is a quasi-autonomous government body (the Sabah Foundation). But entrusting to only one concession the whole of the state's long-term unalienated forest, and that for a period approaching two rotations, appears to be a step too far in the opposite direction. It is suggested that concessions should normally be issued for a complete rotation (sixty years) but that there should be effective ways of cancelling them if work becomes unsatisfactory, and they should only be issued to public limited companies.

Silvicultural systems

Broadly speaking two types of system have been used in the region: the Malayan Uniform System in which the felling cycle corresponds to the length of rotation (approximately 60–70 years) and the next crop depends upon tending the seedling regeneration on the ground at the time of felling; and selective logging systems in which the felling cycle is about half the length of rotation and the next crop depends upon trees of intermediate size which are left standing at the time of logging.

Dipterocarp forest in the region had been traditionally managed on the uniform system except in the Philippines where the Philippines Selective Logging System has been in use since about 1970. But there has recently been a general swing towards selection working throughout the region. There are several reasons for this change: many of the trees girdled under the Malayan Uniform System were found to have become merchantable within a few years; the standard for acceptable logs became lower as forests were worked out; it was found very difficult to delay fellings if regeneration was inadequate, as required by the Malaysian Uniform System; the dense undergrowth after felling led to unsatisfactory recruitment of new seedlings; and above all selection fellings offered an enhanced harvest because of the approximate halving of the felling cycle.

In Peninsular Malaysia the system used now in all states is called the Selective Management System (SMS). This name is unfortunately liable to misunderstanding since the word "selective" has nothing to do with the selection of trees to be felled. It simply means that, after inventory, an appropriate silvicultural system is selected for each compartment – the uniform system, clear felling and planting, or a selection system with different girth limits for different groups of species. In fact it is a means of adapting the system used to the conditions in the individual stand.

In Sarawak selection fellings are used, with a diameter limit of 48 cm and a felling cycle of twenty-five years. In Indonesia the Indonesian Selection System is used. In Papua-New Guinea there is a diameter limit of 50 cm but no clear silvicultural system in use.

In Sabah the only long-term operator in permanent forest, the Sabah Foundation, is still operating under the uniform system with a girth limit of 6 ft (58.2 cm diameter). The foundation has resisted attempts to convert to selection working since this would almost certainly result in large areas being relogged at a lower diameter limit: this would cause extensive damage to the developing stand.

The cutting limits and felling cycles in use within the region are generally similar, as shown (somewhat simplified) in Table 5.10.

Selection working with diameter limits of 45 to 50 cm and a cycle of 30 to 35 years is thus common, and these fellings rely on the relic stand of trees of 20–50 cm diameter to produce the next cut. The diameter increment of all marketable species in Peninsular Malaysia is perhaps about 0.8-1.0 cm per year. After a 30-year cycle, trees of 30 cm diameter could be expected to have reached 54 cm, and trees of 45 cm to have reached at least 69 cm. These will certainly be acceptable commercial sizes in 30 years. So long, therefore, as sufficient relict trees of acceptable species, form, health and freedom from logging damage are left, the system would appear to be operable at least for the first two felling cycles, though more research is needed into the long-term effects.

Table 5.10: Cutting limits and felling cycles in the Asian region

Country	Silvicultural system	Cutting cycle (years)	Diameter limit (cm)
Malaysia			
Peninsular	uniform	55–60	45
	selection	30	50 dipterocarps
			45 non-dipterocarps
Sabah	uniform	60	58
	selection	35	50
Sarawak	selection	25	48
Philippines	selection	30-45	(retain:
			70% of 15–65cm class
			40% of 65–75cm class)
Papua-New Guinea	selection	?	50
Indonesia	selection	35	50 but jelutong 60, *Agathis* 65

It must be said, however, that examination of pre-felling inventories in Indonesia suggests that only rarely does the stand contain sufficient putative residuals of commercial species to enable a further cut to be taken in thirty-five years. A detailed review of the Indonesian Selection System (TPI) is shortly to be made to examine this and other aspects of its application.

It is, in fact, difficult to convince politicians that there may be disadvantages in the doubling of the area under exploitation that accompanies a halving of the cutting cycle, particularly when those disadvantages can be deferred for thirty years. As a result, there is a strong temptation to accept the enhanced coupe associated with adopting the selection system, and to gloss over the fact that it must be supported by a truly viable relict stand of really acceptable, fast-growing, potentially timber-sized, well-formed, healthy trees if the second and subsequent cuts are not to prove a disastrous disappointment.

Pre-felling inventories

Pre-logging inventories are carried out in Peninsular Malaysia, the Philippines, Thailand (but mainly to determine the yield), Sarawak (but only to assess coupe value), Papua-New Guinea (to equalize yield over period of agreement) and sometimes Indonesia. Only in Peninsular Malaysia and the Philippines is a detailed inventory made of trees below the felling limit so that a proportion may be retained undamaged for future cycles, though theoretically such an inventory is also an essential feature of the Indonesian Selection System.

Marking for fellings

In Thailand trees for felling are marked by the Royal Forest Department, the direction of fall being prescribed. This applies rigidly in the teak forests, but in the dipterocarp forests of Peninsular Thailand, where much forest is being cleared for alienation, no marking is done. Directional marking is said to be done rigorously in permanent forest in Peninsular Malaysia, but not in Sabah or Sarawak.

In the Philippines there are detailed marking rules for fellings under the Selective Logging System, trees for retention being marked as well as those for felling and the latter being marked with direction of fall. But practice does not always follow theory and it is not uncommon to mark trees for retention after exploitation, thus avoiding any penalties for felling damage. Markings are carried out by tree markers belonging to the company under the supervision of the Forest Department timber management officer, a supervision, however, which may in many cases be little more than nominal. Also trees for felling may be marked on the butt rather than the stump and the directional felling arrow be placed only on the butt log. Thus, once the tree has been yarded, there is no evidence whether its working complied with the marking. Fellings are also sometimes marked only with paint, no secure marking hammer being used. It is then possible for the company staff to mark up the stumps after felling to make sure that they comply with theoretical felling rules. Two hammer marks, one on the stump directly opposite the line of fall and the other on the butt log, make the marking fool-proof. These marks should be supplemented by a serial number and a paint ring visible from a distance.

No marking for felling is carried out in Papua-New Guinea nor is marking universally applied in Indonesia, though some companies mark voluntarily to control their fellers; the marking is not directional.

Control of exploitation damage and quality of residuals

The swing towards selection fellings throughout the region has already been noted, a felling cycle of one-third or one-half of the rotation being prescribed; thus advantage can be taken of pole-sized advance growth in the original stand. The residuals so retained must be of acceptable species, healthy, undamaged and well distributed.

Tropical rain forest consists of a mosaic of stands of widely varying age. The density of immature residuals also varies widely through a compartment. The incidence of catastrophic damage to primary forest by wind, large-scale drought and fire, mass movement of soils on hillsides, earth tremors and other factors is now thought to have been much greater than was previously supposed and may well explain the presence of dense stands of poles in some areas and the presence of groves of very large emergent trees in others. This variation makes it difficult generally to assess the adequacy or otherwise of relict stands throughout the region without a long period in the field. Some twenty-five years of experience in the dipterocarp forests of Malaysia lead this author, however, to have doubts about the adequacy of the residual stand in many areas.

The best residual stands of the faster-growing dipterocarps probably occur in the drier parts of the region, notably eastern Sabah and the Philippines, where the climate is not so dry as to support only apitong (*Dipterocarpus grandiflorus*) and where *Parashorea plicata* is present. The residual stand in Peninsular Malaysia is locally rich in pole-sized dipterocarps, particularly in lowland forest where *Shorea leprosula* and *S. parvifolia* are common. Kapur (*Dryobalanops aromatica*) also has a good stock of pole-sized trees. It is reported that *Anisoptera thurifera* forms well-stocked pole stands in Papua-New Guinea.

In the steep hill forests of Peninsular Malaysia the ridges, particularly in granite country, tend to be capped with heavy stands of over-mature *Shorea curtisii* with few intermediate-sized dipterocarps, while the middle and lower slopes of the hills and the valley bottoms carry very few dipterocarps of any size. The reasons for this are obscure, but mass movement of the hillsides, periodic water deficits on the ridge crests, and invasion by the stemless palm *Eugeissona tristis*, which inhibits dipterocarp regeneration, all appear to be involved. Apart from dipterocarps, the relict stand will consist of species that are not very attractive to the timber trade: some species which are very slow growing and have hard wood, such as *Eugenia*; some which are of good form and reasonably fast growing but rarely reach timber size, such as *Calophyllum*; some which have properties which make them normally unexploitable, such as *Melanorrhoea*; as well as many which are just about passable as timber species, such as *Endospermum, Scaphium, Lithocarpus* and others.

It will always be necessary to exercise strict control over exploitation damage if forests are to be worked on a poly-cyclic system. Strict standards must be enforced as to which species are acceptable, the amount of damage which can be tolerated, and the crown development and clear bole length which can reasonably be expected to produce a merchantable tree in the next cycle. Where fellings must proceed for economic reasons and the residual stand is unacceptably poor, compartments can be regenerated by transferring to the uniform system, but the cycle for these areas must be extended to sixty years and yield control adjusted accordingly.

Yarding: methods and slope limits

Yarding is the term used to describe the extraction of logs or tree lengths from the stump to a landing where they are loaded on to road-haulage trucks or in rare cases for transfer to a rafting stream or a forest railway. Logs may be moved by animal (buffalo or elephant), by hand-hauling on pole skidways in peat swamps, by caterpillar tractor, wheeled skidder, winch-truck or by cable-yarding to a spar tree.

Animal hauling does the least damage to the residual stand and to the ground surface but is not normally practicable in large-scale operations where considerable volumes of relatively low-priced logs must be moved. It is still used, however, in the very valuable teak forests of Thailand and in Sri Lanka. Hand-hauling also does little damage, though control must be exercised over the cutting of the large quantities of poles used for the skidways. Wheeled skidders, winch-trucks and caterpillar tractors cause relatively more damage to the ground surface because skid trails are needed for them to operate, but if reasonable control is exercised, they do relatively little damage to the residual stand.

Cable-yarding reduces the damage to the ground surface, particularly if skyline is used so that logs may be swung from one spar tree to another without the need to build roads between the spars. But it can be extremely destructive to the relict stand. Only in the Philippines has real progress been made in reducing such damage by limiting the horsepower of yarders and by using rub-trees and snatch blocks to deflect the logs from valuable trees. If cable-yarding is to be permitted in selection working, very experienced operators are required, forest department control must be rigorous and unquestioned over long periods, yarder horsepower must be limited to about 150, cableways limited to about seven per setting, and rub-trees, extended chokers and snatch blocks used as routine procedure. In the Philippines, for example, it will only be possible to enforce these conditions in very few concessions, and it is accordingly recommended that as a general rule cable-yarding should not be permitted in selection workings. Nevertheless, the high stocking of intermediate-sized dipterocarps, which occurs in many forests in the Philippines and in some forests in Sabah, might enable cable-yarding to be used where it would otherwise be unacceptable. Also, where fellings are carried out under the uniform system with regeneration from seedlings and saplings, there can be no objection to cable-yarding, particularly since it tends, if properly controlled, to do less ground damage than tractor working. High lead is now banned in Indonesia and, according to the Forestry Department, also in Sabah (although it is still in use at Silam).

Slopes steeper than 30 degrees should generally not be logged, although it is difficult to exclude all such areas from coupes. In fact not all agreements include a slope limit but extensive areas of steep slopes are usually excluded.

Transport systems, road design and maintenance

Almost all concession agreements have provision for control over the location of main roads and in many cases of spur roads as well. Some forest departments (e.g. in Peninsular Malaysia and Sarawak) have engineers to control this work. Main road alignments seem to be generally satisfactory. Provided these are controlled, it hardly seems necessary to exercise control over the location of spur roads. It would be better for forest departments to ensure that skid trails are properly located and planned before felling takes place in order to reduce as far as possible the damage that might be caused by road and trail building and by yarding tractors.

Main roads give access to shifting cultivators and illegal loggers. In Thailand there is provision to oblige lessees to prevent access along closed roads by rendering them impassable, but this is rarely if ever enforced. There may well be a case for such action in special cases. For example, the removal of one bridge on the road between Sapulut and Keningau in Sabah would effectively prevent the entry of shifting cultivators from the west coast to the whole of the Sapulut-Kalabaken road system.

But, in general, it is better to keep roads open to enable maintenance which will limit erosion and permit sampling and silvicultural work. This requires road formation of an acceptable standard with permanent or semi-permanent bridges and culverts. The maintenance of main and spur roads is very often inadequate, the worst features being the ponding of water caused by blocked or inadequate culverts, the lack of adequate drainage (particularly the lack of cross drains resulting in large volumes of water accumulating on the inner side of roads and scouring the drains very deeply), the lack of check dams to reduce the velocity of water in drains, the lack of protective works to prevent erosion where cross drains and culverts discharge water on steep hillsides, and erosion of bridge abutments.

Where roads and skid trails are abandoned after logging, severe erosion frequently takes place, particularly in steep granite hills. The responsibility for erosion control in such circumstances should be defined; the concessionaire normally has plant in the area and can carry out the necessary work most expeditiously. The possibility of seeding abandoned road formations with grass or creeper seed in a bitumen sealant should be examined.

Post-felling inventories

Inventories after felling should be a normal part of good forest management but they are rarely carried out. In Thailand intensive inventory is carried out only in the valuable teak forests of the north. In Peninsular Malaysia it has long been a routine operation in permanent forests but in Sarawak there is no post-exploitation inventory, nor is there in Sabah. In the Philippines it is routine, though as indicated earlier, it is sometimes combined, irregularly, with a marking of residuals for retention. There is no post-exploitation inventory in Papua-New Guinea or in Indonesia.

Planting voids and enrichment planting

Much is often made of planting landings, roadsides, camp sites, etc. with fast-growing species. A great deal of this work is purely cosmetic and, as seen at Pagdanan in Palawan, Philippines, for example, it can prove to be counter-productive since the trees, which are usually *Albizia falcataria*, shade the roads, keeping them wet and thus easily rutted, and the large soft-wooded trees tend to be blown down or broken by wind, causing a danger to traffic. If voids are to be planted, it would be better to use a valuable hardwood, such as *Swietenia* (in places where the shoot borer which often affects it is not a pest), *Pterocarpus* or even the softwood *Araucaria*.

But the question, rather, is whether voids should be planted at all. This depends on the nature of the site. Landings, road formation and the areas around spar trees on high lead settings are frequently so compacted that ripping or subsoiling with a crawler tractor is essential before planting. As many of these sites will be used once again for the purposes for which they were originally cleared, it would often seem pointless to plant them. Enrichment planting of poorly stocked forest was at one time standard procedure in many operations in Peninsular Malaysia, the work being done by the Forest Department using dipterocarp wildings (wild seedlings collected in the field). Such planting is very difficult to supervise; and the plants must be continually released from competition if they are to survive in the regrowth which follows logging. In general it would be better to invest money in compensatory plantations of genetically improved material of valuable species in relatively easy and accessible sites than to disperse effort over vast areas of steep and inaccessible hillsides.

Security of worked coupes

In five countries (Thailand, Sarawak, the Philippines, Papua-New Guinea and Indonesia) logged-over coupes were often found to have been partially cleared by shifting cultivators. In most cases the situation had political overtones. Frequently the courts are unprepared to take effective action when cases are brought before them. In Mindanao (Philippines) the New People's Army, a communist insurgent group, has threatened to kill forest staff who interfere with illegal cultivators. In fact it does not seem that pressure of population and landlessness are the only reasons for illegal cultivation in forest concessions; the situation could almost certainly be improved by properly organized settlement, with issue of land title, into areas of alienable land. There seems to be no real reason why concessionaires' roads should be deemed to be public roads when they go nowhere except to the forest reserve. This matter should be examined with a view to restricting access to concessions, as is done in some parts of the Philippines.

A further development, seen in Sumatra, is to license part of each concession for industrial tree plantation and then plant a band of about 1 km wide along both sides of the main access road and any spurs which cannot be blocked off. This appears to deter shifting cultivators and, moreover, provides them with useful employment. A similar system has been used by the Paper Industries Corporation of the Philippines (PICOP) at Bislig, Mindanao, though it is not clear whether this was in fact designed as a way of restricting shifting cultivation.

As already mentioned, the protection of concessions in the Philippines is the responsibility of the concessionaire, and in theory their coupe is reduced if they allow forest to be destroyed. In Sabah, on the other hand, worked-over coupes cease to be part of the concession and thus enjoy no protection from the concessionaire. This is an unfortunate arrangement which should be changed.

It cannot be overemphasized that the illegal, and perhaps sometimes the legal, clearing of logged-over coupes in forest which is in legal terms "permanent" is, after the over-issue of felling licences, quite the most powerful factor leading to the non-sustainable management of the moist tropical forests of the region. It is perhaps at its worst in the north and east of Thailand where the rather drier conditions make the forest particularly vulnerable to destruction by fire. But it is also very serious in the Philippines, Sarawak and Indonesia, and it could easily become serious in Sabah. Forestry departments cannot tackle this problem effectively on their own. Among others the following factors require attention before the situation can be brought under control:

- there must be adequate land outside the permanent forests for the existing and likely future population, and increase in numbers should be under control
- there must be one or more profitable permanent crop suitable for the area
- there must be an efficient land administration, including a survey and settlement organization and an agricultural extension service
- the permanent forests must be constituted after a proper public inquiry, and then be properly surveyed and the boundaries clearly marked
- the forestry departments must be adequately staffed and have adequate transport to enable vulnerable boundaries to be patrolled regularly
- there must be the political will to preserve the forest, and the courts must support the forestry departments
- there must be severe penalties for concessionaires who condone illegal settlement within their concessions
- terrorism and insurgency must be controlled.

It is, of course idle to pretend that these conditions are met, or indeed can be met in the foreseeable future, in many of the permanent forest areas of the region. One must then consider whether the sustainable potential of the forests should be defined as only that which the forestry departments can hope to control effectively. The corollary would be that forests in particularly lawless areas and forests which are particularly vulnerable to destruction should be excluded from the productive forest area and be regarded as purely protective. But even this would only be effective if resources were made available for their regeneration. The concept of a random audit of regeneration by an organization remote from the day-to-day administration of the forests seems to be an excellent one. The nucleus for such an organization already exists in the Forest

Management Division of the Peninsular Malaysian Forestry Department, and similar arrangements could no doubt be made in other countries, the team reporting directly to the Director General of Forestry or to the Minister for Forestry.

Recruitment of the regeneration stock

The following comments apply only to dipterocarp forests. The regeneration stock of these ebbs and flows with seed years which occur at intervals of five to seven, or even more, years. Seedlings of most fast-growing dipterocarps only survive for about five years under the dense shade of mature forest. It follows that there are periods under a regular felling regime when felling takes place over inadequate regeneration. This of course is one reason why the uniform system is out of favour. If selection felling is practised, however, the relict stand will be sufficiently clear below for regeneration of dipterocarps to take place within perhaps ten years of felling. Thus on a thirty-year cycle there should be up to three opportunities for substantial recruitment by seedlings, so long as seed trees are left standing when felling takes place. If, after the first cycle, the recruitment of seedlings and saplings is inadequate, it may be necessary to release regeneration by cleaning operations between fellings. It has not been possible to make a study of the situation in the forests of Papua-New Guinea.

Monitoring of the developing stand

The use of permanent sample plots within the exploited forest is an excellent way of monitoring the developing stand, such plots being treated in exactly the same way as the stand which they represent. Such plots are used in Peninsular Malaysia and there was an extensive series of them in the Philippines which has, unfortunately, been discontinued. Some of these have been cut over by shifting cultivators and, generally, there were more plots than staff to maintain them. When new plots are planned, care should be taken that the scheme is not too ambitious.

EXTENT OF SUSTAINABLE MANAGEMENT IN THE REGION

With the exception of Papua-New Guinea where the land tenure system makes sustained management almost impossible, almost all the forests under concession agreement within the region are at least nominally under management. This management includes planning and control as follows:

- yield or coupe should be confined to prescribed limits
- fellings should be orderly and complete, and confined to coupe boundaries
- the residual stand should be adequate
- the residual stand should be protected
- silvicultural work should be carried out
- a felling cycle should be followed, with relogging not permitted between cycles
- permanent roads should be maintained and post-felling erosion controlled
- unworked forest should be protected
- working plans should be written and enforced.

Management tends to be inadequate in all but the first three of these items, although in some countries the others too are adequately controlled. The most serious shortcomings are failure to protect the unworked forest and the residual stand. Sarawak, Thailand, the Philippines and parts of Indonesia have room for improvement in this respect.

Erosion damage to roads and worked coupes is light in northern Thailand, due to the use of elephants and the low rainfall, but over almost the whole of the rest of the region stricter control is needed.

Silvicultural work appears to be taken seriously only in Peninsular Malaysia and possibly in parts of Indonesia. In Sarawak the system of liberation thinnings devised by the Sarawak-FAO team has been largely abandoned as too complicated for practical application.

Relogging between cycles has taken place in Peninsular Malaysia and the Philippines and there is strong pressure for it to be permitted in Sabah. In the Philippines the foremost long-term operator, PICOP, has applied for its felling cycle to be reduced from thirty to twenty years. The writing and enforcement of working plans is in its infancy in most countries in the region but, in general, the concession agreements contain adequate working rules to control fellings on a sustained basis. Adequate enforcement of a few simple rules is far more useful than the proliferation of complex working plans and mathematical yield control methods accompanied by inadequate enforcement.

During this tour only a minute fraction of the operations could be seen and this author was probably taken to see the best. Of these, the following forests and operations appeared to be reasonably success-ful as potential sustained-yield units: in Thailand, the Mae Poong forest; in Peninsular Malaysia there are many possibilities (Jengai Forest Reserve in Trengganu is one of the earliest examples of a forest worked under the selective management system); in Sabah, the operations of the Sabah Foundation, but as uniform rather than selection working; in the Philippines, PICOP at Bislig in Mindanao, but this huge operation has turned from natural-forest management to fast-growing industrial plantations of *Albizia falcataria* and *Euca-lyptus deglupta* to supply its factory – Nasipit Lumber Co. at Butuan is perhaps a better example; in Indonesia, Padeco, Sungei Rawas, and the Musi River, Sumatra.

There is clearly great scope for improvement, and indeed many countries are taking the appropriate action. In Thailand, for exam-ple, there is the establishment of forest development units where all aspects of forestry are represented, but efforts are concentrated in relatively small areas of good forest. It is probable that the Royal Forest Department will only regain control over its forests if it concentrates its efforts on the best areas and surrenders the degraded forests for alienation. The Director General is in favour of concentrating operations by planting high-yielding clones of teak on relatively small areas to replace the logged-over and often degraded coupes. In view of the intense pressure from illegal cultivators and loggers, this seems a reasonable policy but it will inevitably entail loss of some forest land to cultivators.

In Peninsular Malaysia there is a concerted effort at the federal level to codify and enforce the selective management system in the states.

In Sabah the policy of the present director is to divide the perma-nent forests between the Sabah Foundation in the east and south, SAFODA (at present a plantation authority) in the north and Sabah Forest Industries (who own the Sipitang Pulp and Paper Mill and a large block of forest in the Lumaku area to supply it) in the west. The industrial authorities concerned are not universally anxious to assume these responsibilities, which some see as an abdication of central authority. It is clear that an intense battle will soon take place to secure the relogging rights over the huge area of forest which has been logged to a 6 ft girth limit (58.2 cm dbh), much of it in the days when only the highest grade logs were saleable for export and when even trees of the genera *Shorea*, *Dipterocarpus* and *Dryobalanops* were not cut if they were "sinkers". There is also certain to be increasing pressure by shifting cultivators from the west coast now that roads link the heavily populated west with the still-forested east. It is understandable that the director should wish to distance himself from these matters which are sure to be largely a question of politics, but it is less certain whether that is where his duty lies as the custodian of sustained-yield forest management in the state.

In the Philippines the replacement of existing concession agreements by timber production sharing agreements (TPSAs) is a major development now operating on a trial basis. In these government will have the major share and current stumpage rates will be charged on logs, raising their cost at stump to 300 pesos per cu. metre from 30 pesos. There seems to be little advantage in this arrangement as far as sustained-yield management is concerned, but it is an inevitable reaction against concessionaires who have for so long made enormous profits from the Philippine forests and have done so little to perpetuate them. The Forest Department is now being decentralized so that provincial and district foresters will be answerable to provincial councils under the general supervision of regional directors stationed in the provinces. The Director General will still have line responsibility but this will shortly disappear. It is highly questionable whether it is wise to decentralize forest departments at the present juncture; for it will almost inevitably lead to further destruction of the forest and to a decline in sustained-yield management.

Adequacy and independence of staff

Staffing levels in most forestry departments in the region have risen in recent years as more detailed professional forest management has been undertaken; the department in Peninsular Malaysia, for example, employed 5,741 people in 1986 and was the largest employer in the government sector. The department in Thailand had, in 1985, 6,829 forest officials and 4,309 permanent employees. In general, revenues from the forests have been so large during the last decade that governments have been reasonably liberal with staff. Only in Papua-New Guinea does there appear to be a serious shortage, this being due as much to difficulty of recruitment as to lack of funds.

It is difficult in a short tour to form any consistent impression of the efficiency with which exploitation is being monitored and controlled by forest department staff. Nevertheless, it appears that the smaller the operation, the more thoroughly it is supervised. This is partly because concessionaires usually have their own supervisory staff in larger operations and are thought, not perhaps always correctly, to have the long-term interests of the forest at heart; but it is also because most large operators deliberately discourage forest field staff from investigating too closely the conduct of their operations.

Lack of government support (in providing housing, transport, schooling, medical facilities, food supplies, electricity and water) makes forest staff largely dependent on the goodwill of concession staff, and this situation is often exploited to the full. If necessary, bribery may also be used to persuade staff to overlook breaches of the forest law and, in one case in the Philippines, forest staff were forbidden to enter the concession without the approval of the concessionaire.

Political influences, too, make it difficult for forest department staff to examine operations very closely. This situation will continue as long as politicians intercede with government on the behalf of concessionaires, political figures achieve positions of influence in logging companies, and senior administrative officers of government become involved in matters which are the simple concern of the forest department. In Thailand it is reported that the forest staff do not find the Forest Industries Organization easy to deal with, simply because its chairman and board members are far too powerful. In the Philippines almost all large logging companies have very senior politicians on their boards and the same is true in Sabah. In Papua-New Guinea political influence in the running of logging and timber export operations was recently the subject of an official public inquiry. In Peninsular Malaysia logging companies tend to be smaller, but of recent years large semi-government organizations have entered the field and these cannot be easy for forest officers to deal with. In Indonesia there are some very large local companies involved in the timber industry and it is said that some of these do not hesitate to bring pressure to bear in Jakarta if local foresters are too insistent in investigating breaches of concession agreements.

Apart from the obvious need to remove political influence, the situation could be greatly improved by making forest staff as independent as possible of concession staff in logging camps; certainly they should not live in company houses or use company transport. In addition there should be a forester of senior rank in every concession area who can meet the company manager on equal terms; this officer should have wide powers delegated to him. Care should be taken to post officers with strong personalities to difficult concession posts.

Funding of forestry departments

The resources available to forestry departments have expanded enormously during the last decade and, with the possible exception of Papua-New Guinea, there is no evidence that facilities are inadequate. Senior officers of the Royal Thai Forest Department and of the forestry departments of Peninsular Malaysia, Sarawak, Sabah, the Philippines and Indonesia have become highly professional; for example, the most sophisticated mapping and computing equipment is available to them and most of these departments have head offices which can only be described as lavish.

Decentralization of forest departments: state versus federal forests

In Peninsular Malaysia forests and land have long been state, not federal, subjects and forest revenue has been state revenue. Sabah and Sarawak, similarly, have independent forest departments. The Philippines, too, is now decentralizing its forest department. Thailand and Indonesia alone have remained highly centralized. Assuming that the headquarters of the department is staffed by the most experienced and competent officers, it would seem to be desirable that the centre should control management of the national forest estate. Forests are extremely valuable and their destruction is likely to bring enormous short-term gains to the destroyer. The worst excesses of precipitate action are more likely to be moderated if the disposal of forests and their exploitation and silvicultural treatment are monitored by an organization somewhat remote from the forest itself and from those who would profit by its exploitation or destruction. Any proposals to alienate permanent forests should certainly be examined at headquarters and there should also be a central audit of forestry regeneration operations.

This point can be illustrated by Malaysian experience. In the past there has been heavy pressure to open too much forest to exploitation since this could enable states to acquire funds for state projects which would be unlikely to qualify for federal finance. Similarly roads could be built by timber licensees in exchange for forest licences; and facilities such as schools and rest houses could also be built in exchange for logging rights. A National Forestry Council (NFC) was formed in Malaysia in 1971, chaired by the Deputy Prime Minister, to co-ordinate forestry policy and practice. All concession coupes and the issue of new concessions now require the approval of this council. This is an important step towards controlling the opening of forest to exploitation.

Research and training

The region is well supplied with forest research institutes, many of them established for a long time and with world-wide reputations.

In Thailand there is the Forest Herbarium, Bangkok. There is also a new silvicultural research institute, a forest products research laboratory, and inventory and management departments.

In Peninsular Malaysia the Forest Research Institute has now become an independent organization separate from the Forestry Department, and is known as the Forest Research Institute, Malaysia (FRIM). This institute pioneered much of the work on dipterocarp silviculture and has a herbarium and wood collection of unique importance. Forest products research is also undertaken.

In Sarawak there is a research branch within the Forest Department at Kuching with an excellent herbarium. Much research on the ecology and silviculture of peat swamp forest has been carried out there.

In Sabah the Forest Research Centre at Sepilok has a herbarium with over 94,000 specimens, a sawmill and laboratories. Work on clonal propagation of plantation species has recently been an important feature of research at Sepilok.

In the Philippines the Forest Research Institute (FORI) at Los Baños is well known and is closely associated with the Forestry Faculty of the University of the Philippines and the Forest Products Research Laboratory (FORPRIDICOM). FORI is independent of the Bureau of Forest Development.

In Indonesia the Forest Research Institute at Bogor is one of the oldest institutions of its kind in the region. It has a herbarium of great importance and the Forest Products Research Laboratory is at the same site. There are many universities in Indonesia with forestry faculties which also conduct forest research.

In Papua-New Guinea the Forest Herbarium at Lae has an

international reputation and collections of great importance. Unfortunately, relatively little work has been published which would contribute to a tree flora of the country suitable for use by foresters. The Forestry Faculty at the University in Lae is active in research. Research on forest products and forest management is carried out at the Forest Department headquarters in Port Moresby.

Forestry training

University training for senior officers is now available in several places in the region. It is no longer the practice to send officers abroad for training, except for certain post-graduate studies. Universities offering degree courses in forestry include the Kasetsart University of Thailand, the Universiti Pertanian in Peninsular Malaysia, the University of the Philippines at Los Baños and the University of Technology, Lae, in Papua-New Guinea. For studies at diploma level the Forestry College at Bulolo, Papua-New Guinea, and the Forest College, Kepong, Malaysia, offer courses. Field staff are trained at the forest training schools in Sabah and Sarawak, at the ranger school, Phrae, Thailand, and at the forest school, Kepong, Malaysia. In Indonesia there are several universities offering degree courses in forestry and many local forest schools for training field staff. In general the professional officers of the departments are highly trained, many with higher degrees obtained in overseas universities. The standard of training of the field staff is also impressive.

However, there is one serious deficiency. The standard of tree and timber identification by senior officers in the Malaysian states, the Philippines and Indonesia leaves room for improvement. As more and more species become utilizable, it will be necessary for not only the field staff but also senior officers to be able to identify a much wider range of species in the forest.

PROSPECTS FOR SUSTAINED-YIELD MANAGEMENT WITHIN THE REGION

Sustainable supply

The out-turn of timber from forests in the countries of the region has fluctuated widely in the past, partly because a very large part of the production has come from the liquidation of the accumulated capital of conversion forests when they were alienated, at least nominally, to agriculture. Log exports from the region peaked at 39.1 million cu. metres in 1973 and declined to 18.7 million cu. metres in 1982, although the total was also controlled from 1981 onwards by the imposition of increasing restrictions on the export of logs.

Sustained supply in the context of this study can only relate to the permanent production forests of the region. A comparison of these, and their approximate sustained-yield potential, with the most recent timber out-turn figures is instructive.

Table 5.11: Timber production and yield potential in the Asian region

Country	Permanent production forest (million ha)	Sustained yield at 1.5 cu. metres/yr	Sustained yield at 2.0 cu. metres/yr	Annual timber out-turn (cu. metres)
Malaysia				
Peninsular	2.85	4,275,000	5,700,000	7,914,328
Sabah	3.00	4,500,000	6,000,000	11,739,262
Sarawak	3.24	4,860,000	6,480,000	11,470,000
Philippines	4.40	6,600,000	8,800,000	3,433,774
Indonesia	33.87	50,805,000	67,740,000	28,500,000
Total	**47.36**	**71,040,000**	**94,720,000**	**63,057,364**

In Table 5.11 the approximate sustained-yield potential of the permanent production forests in the region is compared with the most recent timber out-turn figures. It appears that Sabah and Sarawak are still liquidating substantial forests for alienation (or overcutting the permanent forests); Peninsular Malaysia is approaching a balance, but still liquidating some forests for agriculture. The Philippines is having difficulty in finding enough workable forests to satisfy the coupe; or else the productive forest area has been overestimated; or both. Indonesia either has considerable scope to expand sustained production, or has overestimated the extent of productive forest, or both.

In the future the quality of timber produced from the forests in the region is certain to change. It will soon cease to consist of high grade dipterocarp logs with an average volume of 3-4 cu. metres, lengths of up to 11 metres and mid-girths of up to 3.6 metres or even more. The out-turn of the future will consist of logs with maximum lengths of 5 metres, diameters of about 30 cm; it will no longer be free of defect. Moreover there will be many more species, perhaps 200-300, for there are some 700 possible merchantable species in Peninsular Malaysia alone.

The main factors hindering or preventing sustained-yield management

These vary from country to country but illegal clearing of forests, and particularly of worked-over coupes by shifting cultivators, is by far the most damaging factor in Thailand, the Philippines, Sarawak, Indonesia and, potentially, parts of Sabah. Parallel with this is illegal logging of the unworked forest and of residual trees in worked-over coupes; this is common in Thailand and also in the Philippines. The provision of road access for logging is a strong contributory factor which can be countered by the closing of forest roads after logging, the destruction of bridges, trenching and provision of guard points.

The second most damaging practice is allowing forest to be exploited either in excess of the sustainable cut or when it is silviculturally damaging to do so. Professional foresters do not have the power to restrict fellings as they would wish under regimes where individual politicians can override professional advice and license the opening of forests for their own political or pecuniary advantage. It would represent real progress if some other countries in the region were to follow the example of Malaysia quoted above (p. 148). However strongly professional foresters may feel that the opening and operating of forests should be within their control, if a democratically elected government were to direct that all the forests in the country be destroyed and replaced by shifting cultivation, there would be nothing that they could do except to advise against it. In these circumstances the forester clearly has a duty to educate the population on the benefits of forest conservation and sustained-yield management. The head of the forest service also has the duty to argue the case with ministers and must be sufficiently senior to be able to do this with authority.

The techniques of regenerating and managing tropical moist forests on a sustained-yield basis are now at least well enough understood to enable sustainable management to proceed, and in general forestry departments are well enough staffed and funded to do so. But this also requires sufficient well-trained staff to be deployed in the field. Some countries have problems in doing this; in Indonesia, for example, more than half of the trained professional foresters are stationed in Jakarta.

Good conditions of service and field incentives will do much to resolve this situation: the provision in concession areas of good, even lavish, housing, schools, facilities for recreation, medical facilities, adequate transport, and arrangements for leave (and transport to take it); it should be required that staff have their families with them and do not merely camp in concessions. The amount of funding required would be small compared with forest revenue, and minute compared with the long-term value of the forest. The situation requires attention in the Philippines and Indonesia now, and it will shortly do so in Papua-New Guinea.

The status of forest officers and their subordinate staff in concessions is also a matter of concern, particularly in the Philippines. Certainly, provision of adequate housing and transport will go a long way to improve the status of officers. But it is also important that the forest officer stationed in each concession should be of high enough rank to meet the manager on equal terms. In no way should the forest staff be forced to rely on concession staff for transport, water, electricity, housing or any other essential of life; and there should be a proper government resthouse for visiting government officers.

Circumstances where sustained-yield management is succeeding

The most promising area in the region for sustained yield management is Peninsular Malaysia. Here, a long-established forestry department with a high reputation, well-trained and disciplined field staff, excellent research facilities and a reputation for guarding the forest and resisting excisions, even to the point of stubbornness, has been a powerful factor in its success. There has also been a strong tradition of managing the forest in the field and not in the office. The removal of licensing and regulatory matters from the hands of individual political personages has helped greatly. Perhaps most important of all, the establishment of the National Forestry Council has raised the profile of forestry to the highest level in government.

NOTES

1. The following are key references for the Asian countries:
    Indonesia: Anon., *Pedoman TPI (Guide to the Indonesian Selection System)* (Jakarta: Forestry Department, 1985); GOI/IIED, *Forest Policies in Indonesia, Vols II and III* (London: IIED,

1985); Ross, M.S., *The South Seas Log Market* (Jakarta: Department of Transmigration, 1983).

Malaysia: Anon., *Sarawak Forestry Statistics* (Kuching: Forest Department, 1986); Anon., *Sabah Forest Department Annual Report* (Sandakan: Forest Department, 1982); Hutchinson, I.D., *Sarawak Liberation Thinning* (Kuching: UNDP/FAO, 1981); Thang, H.C., "Timber supply and domestic demand in Peninsular Malaysia", *Malayan Forester*, 48 (1985); Thang, H.C., *Selective Management System, Concept and Practice* (Kuala Lumpur: Forestry Department, 1987).

Papua-New Guinea: Anon., *Compendium of Statistics* (Port Moresby: Department of Forests, 1986); Anon., *Annual Report* (Port Moresby: Department of Forests, 1987).

Philippines: Anon., *Philippine Forestry Statistics* (Manila: Bureau of Forest Development, 1986).

Thailand: Nalampoon, A., *Thailand Forestry in Brief* (Bangkok: Royal Forest Department, 1986).

2. An alternative, and potentially more effective, system of balancing the claims of forestry against those of agriculture has been proposed by: Ross, M.S. "Forestry in land use policy for Indonesia", University of Oxford, Thesis for the Degree of Doctor of Philosophy (1984); and by IIED/Government of Indonesia, *Forest Policies in Indonesia: the Sustainable Development of Forest Lands* (London: International Institute for Environment and Development, 1985).

3. FAO, Forestry Utilization Contracts on Public Land, FAO Forestry Paper 1 (Rome, 1977).

6. Management of Natural Forest for Sustainable Timber Production: a Commentary[1]
John Palmer

There has been an enormous surge in literature on tropical moist forests in the last decade, including a number of reviews. However, little is new or of direct relevance to management. Most of the literature reports small-scale ecological studies. Some of these studies are components of team efforts to look at samples of ecosystems, but many are unco-ordinated one-offs to provide material for postgraduate theses. The most relevant literature continues to be the now quite old manuals of silviculture and management compiled from the late 1950s to the early 1970s in Africa and South East Asia.

These manuals were largely concerned with a description of the background to and operation of the more successful silvicultural practices, within the particular managerial contexts of the countries for which they were written. In no case were there claims that the described systems had been developed to a definitive state. In some cases the stimulus for a manual was a change in the political system, for example the withdrawal of colonial forest service staff or their immediate post-independence replacements. The manuals were invariably preceded by numbers of separate publications touching on different aspects of forest management. Many of these papers reported intermediate results of research studies.

The managerial contexts were often at those times in a greater state of flux than the silvicultural systems. The reality of management rarely coincided with the national forest policy and this disparity has tended to increase over the last four decades. The reasons for the disparity are social, cultural, economic, political and technical. Quite naturally there has been a reluctance by serving officers in national forest departments to defend in writing the practices which they are required to execute, when these are contrary to the perceived long-term good of the country. Equally, it has been contrary to the long-term practice of international agencies of technical co-operation to describe explicitly the national field practices which are both prevalent and contrary to sustainable land management. Consequently there is a dearth of comprehensive and factual information for all countries having tropical moist forest, since none now practises management of that forest in a demonstrably sustainable manner. Regional reviews which do make criticisms carefully veil them to conceal the identity of the countries with unsustainable management. It is thus not at all surprising that the non-government environmental groups have tended to make unrealistic comments and suggestions, because they have no ready access to reliable and factual information.

This short commentary indicates some of the more important reasons for the present state of management, and what might be done to improve matters. The first part is a diagnosis, the second part is prescriptive.

The situation is not irreparable. Where the forest still exists, good management could be applied now, today. It is not necessary to wait a moment longer for new research to be initiated or published before starting or improving management.

DIAGNOSIS WITH EXAMPLES OF PROBLEMS

Demography

The dramatic increase in world population of course has its roots several generations ago. The gradual cessation of tribal warfare and overt slavery, the advances in public-health measures and the improvement of transport systems have contributed to many societies increasing at near the biological maximum. However, the customs and traditions of human societies have changed relatively slowly. Methods of land management which appeared to be adequate for small populations may be unsustainable for larger populations. This fact is usually quite apparent in any regional or national survey but may be much less evident at the level of the individual farm family or small village. Circumstances which may appear shocking to a visitor, or to an inhabitant returning after a period of absence, may have crept up on a local community, and have caused only minor changes in behaviour, when more drastic changes would have been appropriate.

European schoolchildren may learn about geometric progressions from the story of the pond and the doubling waterweed, but the fish's perception is quite different until the thirtieth day. Temperate-zone anxieties expressed in *The Limits to Growth*,[2] published in 1972, were scarcely mirrored in the developing countries. On the contrary, not a few reacted negatively against what they saw as an attempt to restrict their freedom to "go forth and multiply".

Concerning tropical forest management, the swidden system employed by most Bornean societies was relatively innocuous as long as the populations were small (checked by tribal warfare and disease) and the forest without value outside the immediate community. The land use maps issued in 1970 showed for the first time what enormous loss of commercially valuable forest was caused annually by fresh swidden clearing in Sarawak. Quite apart from any ecological considerations, a state having few natural resources except forest would be entirely justified in placing as much forest as possible in reserves, until a less destructive agricultural method had been accepted and applied by the majority of the rural population. Even after the usual prolonged legal investigations into claims for customary rights and privileges, opposition to reservation has been prolonged and changes to swiddening only minor. In the same country, the Chinese section of the population is not allowed to purchase land in the interior and has farmed intensively on small parcels on the outskirts of towns. The high level of biological and commercial productivity of the Chinese farmers, often on very infertile soil, is well known.

Social factors

In many Brazilian towns, fresh vegetables and fruit are marketed by first- or second-generation Japanese immigrants. Half the family lives outside the town and produces the food, the other half manages the shop or market stall in the town. The families are accustomed to intensive working of small areas of land. They come from a long tradition of land workers and have the knowledge to improve the productivity of the soil. Groups which migrated from southern Brazil to Amazonia in the early 1970s, or from Japan to eastern Amazonia in the early part of this century, are among the few who have made a commercial success of farming the infertile soils, using the same combination of agricultural knowledge and business acumen.

Mestizo colonists in the Amazon have fared less well. They come from a more individualistic society which developed in the Iberian peninsula during the medieval wars with the Moors, and whose attitudes are quite remarkably persistent throughout twentieth-century Latin America. This society praises the male warrior on the horse, the cattle owner, the mine owner. Dreams of becoming a "hidalgo" are more potent, and time-consuming, than compost-making. Whereas the immigrants from Japan, and the colonists settled for generations in the lower Amazon, select carefully the sites for their fields, the new colonists simply take the next unoccupied land. Without knowledge of the soil and the appropriate varieties of the most suitable crops, the new farmers' yields are poor. The value of the forest burned is much the same, whether the crop yields are good or bad. The farmer with the poor crops is more likely to shift and cut another tract of forest than the farmer with the better yield who might stay put.

Similarly, in the eastern lowlands of Bolivia, neither the long-term mestizo Kamba residents nor the immigrant Kolla from the high Andes aspire to a lifetime of farming, even if the reality is that they will indeed be farmers all their lives. Farming is thought of as a means of generating sufficient cash to move into town as truckers or small businessmen. Thus they see no incentive to invest any more than minimal effort into learning how to improve their farms. Much more effort is required to maintain or upgrade the fertility of a tropical farm than to make use of the fleeting fertility offered by the ashes of felled and burned forest. Given their social aspirations, the farmers will always opt to shift their residence and fell more forest, rather than improve what they have now.

Internal migration, from whatever causes, is now the major factor in loss of tropical moist forest. In some countries the driving forces are financial speculation in land and artificial incentives to raise unsustainable crops (such as cattle). The greatest force is the inability to raise a good enough income from the land to make worthwhile a long-term settlement. This inability is sometimes due to social factors (a preference for city life, however uncertain) and sometimes to ignorance of appropriate farming techniques and lack of a financial environment (stable markets, access to credit) to make the effort worthwhile.

Land tenure

In a previous section the Brazilian and Bolivian farmers mentioned enjoyed *de facto* or *de jure* security of tenure over the land which they cleared from forest. The act of clearing is regarded legally in most of Latin America as an "improvement". The maker of the improvement acquires a possessory right and can sell that right. Indeed, for some farmers in Central America, the annual sale of rights to the land cleared that year is by far the largest cash income which the family receives, far greater than income earned from the sale of any excess farm produce. That cash may be a large amount for the family but small change to the buyer. The small cash cost of destroying forest is a major factor in its loss.

In some countries, the colonist who clears forest and then resides on the plot for a period of years may apply for a freehold title; in others, the original clearer cannot acquire title but the person who buys the possessory right from the colonist may do so. The Homestead Acts of the United States have been copied widely in the neotropics, sometimes in attempts to induce farmers to move from overcrowded areas with exhausted soil, and to stem the migration from rural areas to the cities. The existence of the US Acts is sometimes claimed to legitimize such copies, although these operate in quite different circumstances.

Since these possessory rights are sold for a nominal sum, the buyer in effect acquires unencumbered and deforested land for a very low price. However, in order to maintain his ownership, the buyer must use the land. If the buyer has acquired the land for status or speculation, the cheapest method of working the land which is legally recognized is open-range grazing. Although there is the initial cost of fencing, there may be no recurrent costs other than that of a guardian for the cattle. The meat offtake may be an intended output or a bonus, depending on the landowner's aspirations.

The landowner must use the land, or be liable to invasion by squatters. The invaders may be landed or landless families. If they are landed, they may be attempting to add to their holdings or they may be employed by third parties in personal or political feuds. If they are landless, this may be because demographic increase has diminished the available area of farmable land in their home location, or because they have sold their land or the rights to it; they may have been displaced by engineering works such as hydroelectric dams; they may have been tenant farmers whose landlord has taken the farm in hand. In some countries the landless may have been deprived of their holdings by force, perhaps legally because they should not have been there in the first place, or perhaps illegally by the goon squads of expansionist land holders.

In a number of countries, the invasion of privately owned land by squatters is defined in the constitution as a subversive offence. The landowner may be able to call on police and military help to remove the squatters and the government itself may prosecute them with severe penalties. The invasion of public land, such as a national forest reserve, usually carries only minor penalties. Carried to the extreme, as for example in Honduras, the national forests are used in effect by the government as a safety-valve and as a way of postponing the reform of land tenure.

The effect on the forest of these population movements is almost invariably negative. It is clear from the few examples given that security of tenure for farmers in itself does nothing to prevent the clearing of natural forest. However, it is not at all unusual for a farmer to clear the natural forest from his holding and then plant up part of it with fruit or timber trees, sometimes the same species whose mature specimens he removed but a year or two previously. This apparently irrational behaviour seems to be designed first to obtain local acknowledgement of his right of occupancy (by clearing) and second to increase his investment in the long-term good of the farm, once his tenure appears to be secure.

Cultural factors

Even in the areas where shifting cultivation is a relatively stable form of agriculture, the length of the fallow cycle has often decreased markedly within living memory. This reduction may be due partly to the increased planting of perennial cash crops, taking land out of circulation, but mainly to demographic increase. The area farmed annually per family used to be controlled more by the area which could be weeded than by the area which could be cleared. Three factors have contributed to the current substantial increases in the areas cleared: the first is the cheap and reliable chainsaw; the second is the general decline in the trade in minor forest products; the third is the increased market for land, either to spontaneous immigrants or (by way of compensation payments) to government for sponsored settlement schemes.

Apart from groves reserved for the selective harvesting for domestic use of minor products, usually not far from dwellings, there are few cultural restrictions on the area of forest which can be cleared by a family or community. Self-sufficient tribal groups had little incentive to strive for large agricultural outputs, since the markets were too distant and the produce could not stand the freight costs and/or storage losses were high. Items not produced within the group were generally acquired by sale or barter of minor forest produce. Items such as bezoar stones, incense wood and dragon's blood were valued for their magical properties. Gums, latex, oils and resins were valued ingredients in manufactured products before the plastics revolution induced by cheap petroleum. These forest products were often obtained by dreadfully inefficient and destructive methods, though thought to be harmless when the forest was perceived as having no intrinsic value outside the immediate community.

Now these markets have almost entirely disappeared. Rosewood oil for expensive perfumes can be synthesized far more cheaply from the effluent of kraft pulp mills than the natural product can be distilled from wood chips in Amazonia. The few products which have properties which are difficult to reproduce industrially, such as natural rubber, have been taken into agricultural production or are in the process of domestication (such as rattan).

There remains a small group of products which are locally important but not worth synthesizing: examples include durable poles for house construction and agriculture. Finally there are fruits and nuts, canes (rattans) and medicinal plants, some of which are partly domesticated and many are worth further effort. For a developing country, there may be little experience of technological forecasting to determine which of these products are worth greater effort. For example, the rattan market declined markedly during and after the Second World War but is now booming. The research and management efforts being made to take advantage of this market are absurdly small even though the financial return is high. The steadily increasing markets for fruit from tropical moist forest, both within the native countries and for export, have likewise not attracted the private or public investment which would seem to be justified. The failure to invest seems to be connected with the very high rates of return from speculation in property and currency which are obtainable in countries where the overall infrastructure is weak.

There is a currently popular view of the tropical forest as a cornucopia of genetic material. This picture of a vast storehouse of directly usable plants is somewhat mistaken. Certainly, the long-settled indigenous groups make use of a large range of plants for medicinal and cultural purposes, and many of these have been found to have interesting pharmaceutical properties. However, the drug companies have no incentive to rely on the uncertain supply of probably adulterated forest-gathered plants for their expensive medicines. The plants supply the raw material for the isolation of the active principles, and for study of their structures and hints about synthesis. But the drug companies have no interest in stewing cauldrons of rosy periwinkle as a long-term source of pharmaceuticals.

How, then, can the forest-dwellers obtain cash to buy the manufactured products which are needed not only for improvement of the material standard of living but also for the enhancement of the community's social life? One way is through work off-site; groups of Iban from Borneo have traditionally made long journeys in early manhood, often obtaining employment hundreds and thousands of kilometres from home for years at a time. Another way is to participate in the booming tourist industry, to develop craft work and sell it. The most common, however, is to increase agricultural production, not by intensification but by expanding the area cleared and sown, and selling the excess. This is more possible than it was formerly, because transport links are better and cheaper and because burgeoning urban populations provide a larger market than before. Nevertheless, a dependence on low technology and rainfed agriculture often results in seasonal gluts and low prices. The loser, once again, is the forest.

A last point in this section on cultural factors is that the forest supplies different benefits to different people. The time frame of these benefits is least to the logger with the short-term timber licence and greatest to the nomadic forest-dweller. The logger who passes through the forest once and buys hotels in Hongkong with the profit on the sale of raw logs clearly has less long-term interest in the forest than the nomad who is totally reliant on it. No brownie points for guessing whose voice is heard when logging licences are issued.

Technological factors

The effect of the cheap chainsaw has already been mentioned, greatly increasing a family's capacity to clear forest. If they are lucky with their harvest and markets, they may earn a lot. If they are unlucky, their efforts cost little.

The chainsaw has also increased the clearing capacity of immigrants from outside the forest. Some of the land may be sown directly to grass, and cattle may be acquired by various arrangements to graze it. Again, larger urban appetites provide a ready market for any produce while the cost in effort is small.

The chainsaw, the tracked tractor and the logging truck have, of course, transformed the exploitation of timber from tropical moist forest compared with the operations prior to 1940. These increasingly reliable and vastly more powerful machines allow a small number of humans to harvest the largest logs with ease. Logging is a highly profitable operation almost everywhere in the moist tropics. The payback period on equipment is short, the risks are slight. Due to the negligence of the national governments, the raw material costs almost nothing and controls are insignificant or easily bypassed. The most difficult step is often raising the cash to pay the bribes to secure the logging licence.

Economic factors

The value of the forest varies depending on the viewpoint. For the few remaining hunter-gatherer groups, the forest is all; they depend upon its continued existence for everything; they know no other way of life; for them it is priceless. However, they are rarely enfranchised, so governments ignore them when they do not oppress them directly.

The forest-dwellers who practise agriculture and who are in touch with the monetary economy are usually keenly aware of the boundaries between their land and that of neighbouring groups. Solomon Islanders use topographic features, stones and particular trees to differentiate territories. These groups often display no particular enthusiasm for the forest as such, and have been restrained from larger clearing in the past by inadequate technology and tribal warfare. Land reduced to scrub and weed by frequent burning is simply abandoned. Once the soil is exhausted, they move.

Farmers practising a more sedentary agriculture are relatively rare in the moist tropical forest. Land which is inherently fertile is restricted to the occasional base-rich volcanic soils, or to alluvial plains which are refertilized by silt deposition. These areas have often been settled for a long time and natural forest was mainly cleared before the present century. However, forest on river banks may have been deliberately preserved, to pin the banks and diminish the damage caused by flood water. Otherwise, natural forest rich in fruit trees may have been left in groves, and hills may have been left for cultural or religious reasons, or because the soil was notably poorer.

The tropical moist forest now rarely impinges directly on the life of the urban-dweller in developing countries. The forest itself has been cleared for agriculture far from the city boundaries. The logs, trucked in or floated to the urban industries, provide employment, but there is no particular cachet attached to working in a forest industry. The millers complain about the high cost, small size and poor quality of logs which they buy from the logging contractors; the carpenters complain about the defective and badly sawn planks from the mills; the housewife complains about the stench from the vast piles of burning waste, and the fisherman complains about the river pollution caused by dumping of sawdust and bark. The richer urbanites will prefer to buy teak furniture manufactured in Scandinavia and exported to the country of origin of the timber, rather than the products of the local cabinet-makers. Some countries, with wonderful joinery woods, consider that sticky plastic-covered, steel-framed furniture is *à la mode*, preferably imported at vast cost.

The log exporters complain about the atrocious cost of the logs, the cost of shipment, the poor prices and the high export duties.

Apart from the last item, export duties, the values of the forest products at each stage are very largely fixed by market forces. This is not to say that they have to be, rather that the governments have allowed matters to develop in that way. The forest-dweller who relies on water transport will value highly a species which is easy to manufacture into canoes. Most forest-dwellers value species which produce durable house posts. Once these species become annoyingly scarce, through overcutting or loss of forest area, there may be cultural restrictions (taboos) on their cutting, mainly to ensure that the tribal leaders have a continued supply. It is notable that tribal groups which are egalitarian have few restrictions on usage. Some governments ban the commercial logging of indigenous fruit trees and other species of domestic importance or sources of minor forest products. However, compensation payments for illegal cutting are often fixed at absurdly low rates, unchanged for decades.

This is not the place to discuss the various ways in which forest may be valued. However, it is clearly much more difficult to value a potentially renewable natural resource than those which are not renewable (such as mining products), readily renewed (such as farm products) or wholly manufactured. The problem is compounded by the length of time to make that renewal, compared with annual or biennial agricultural crops. A further problem is the valuation of indirect products, such as perennial flows of clean water, stabilization of the soil on steep slopes, and amelioration of micro-climates.

The inability of tropical foresters to suggest ways of valuing the goods and services from the forest, which are meaningful to their colleagues in national treasuries and planning ministries, has been a major factor in the continuing loss of those forests. A discussion of how and why this has come about constitutes the next part of this chapter.

Institutional factors

This section is based mainly on the experiences of the British colonial forest services and their successors. The national forest services in Latin America, which were created long after liberation from Portuguese and Spanish rule, have always operated in a very different ethical environment. There, the concept of government for public good rather than private gain is still not widely appreciated, and perhaps for that reason government services have been relatively unsuccessful. The tragedy of the commons is by no means confined to the neotropics but there the phenomenon is particularly noticeable.

Government-organized forestry had been in progress in much of Europe for well over a century by the time that the Viceroy of India decided, in 1855, that a formal forest policy was necessary for the subcontinent. Concern at the deterioration of the natural teak forests in Pegu (lower Burma), an important source of this strategically important naval timber, led to the recruitment in 1856 of Dietrich Brandis. He was the first of a number of notable German foresters in Burma and India, who had access and influence at the highest levels in government. Historical studies by H.C. Dawkins at Oxford have shown that attempts to manage rain forests go back more than 130 years.[3]

Not surprisingly, the legislative and administrative arrangements made in the latter half of the nineteenth century in India and Burma were used as models in other parts of the British-ruled tropics. A similar process was followed in the francophone territories, which of course drew on the also substantial home-country experience of French government forestry. It is evident from the literature that the greatest concern of the foresters during this period was to slow down or stop the environmental deterioration in the colonial territories. There was a clear appreciation of the effect on population growth and agricultural activity following cessation of slavery and tribal warfare. The indirect benefits of managed forest seem to have been regarded as so evident that the reservation of remaining forest generally enjoyed administrative support, subject to respect for established rights where these were not obviously harmful.

There does not seem to have been much attempt to place financial values on these indirect benefits. Instead, the foresters were able to cite numerous examples of the cash losses sustained through environmental disasters which were attributed to the removal of forest.

At the same time, the impecunious colonial regimes were always keen to see that government services covered the costs of administration through taxation of various sorts. In some areas, where forests contained commercially valuable timbers, the revenue-earning capability of the forest service was much appreciated by governments. It is evident from annual reports that the forest services took pride in being major contributors to national treasuries. An overemphasis on activities associated with revenue collection has opened tropical forest services to the reasonable charges that they have facilitated over-exploitation while being unable to prevent under-utilization and destruction. These charges overlook the fact that politicians in several countries have deliberately weakened forest services so that they cannot fulfil their statutory functions. For example, the creation of politically staffed boards to issue timber licences according to ministerial whims has made management planning almost impossible in much of South East Asia.

In spite of enjoying political support for their activities, no colonial forest service succeeded in pressing its case for sustained funding. The colonial regimes recognized, in the very establishment of the forest services, that the low rate of return and the long payback period made forestry an unattractive private investment. Treasury officials did not accept that this implied any exception to the rule of annual funding, nor did they allow the forest services to retain even a proportion of taxation to establish a long-term development fund. There were a few exceptions to these rules, but they were all short-lived.

It is this author's view that forest services should have spent and should now spend much more effort on exploring fiscal alternatives. Such efforts could be along several lines.

First, they should attempt to assign values to forests which are understood and accepted by funding agencies (national or otherwise), and make realistic estimates of the costs of sustaining and improving forest benefits, both goods and services.

Second, many forest services have devoted a large amount of effort to improving the collection of minor taxes. Although this was the *raison d'être* of most Latin American services (at least initially), in the old-world tropics the forest legislation usually did not require that the forest service itself should play more than a minor part in tax collection; this was an obligation taken on by the services, but for which they were never suitably staffed, trained or equipped. It can be argued that foresters should have confined their revenue collection to the provision of data to the national treasuries or tax authorities and spent instead much more effort on the forest management for which they were trained. If the foresters are unwilling or unable to shed their current role as unpopular and inefficient collectors of revenue, they should at least make a better attempt to devise systems which are less open to bribery and corruption. For a start, the taxation rates should be placed on a more logical basis and related to the real cost of supervising logging concessions and other management activities required to maintain or improve the sustained yield of the forest. For example, a management charge per unit area of logging licence would be easy to calculate and simple to administer. Undoubtedly, any rationalization of current systems will upset some people who benefit from the present situations. Unfortunately, many previous attempts to improve systems have failed, and their authors have been subjected to various kinds of unpleasant censure: loss of pension rights; removal of citizenship; prosecution on fabricated charges.

A more subtle approach should constitute the third line of effort. As central governments become increasingly unable to finance services which have no direct political benefit, the forest services should be exploring different ways of charging for their work. This means looking at the way other branches of government justify their budgets. It means a more considered approach, with greater awareness of the self-interest of decision-makers. Obviously, this effort would depend on developments in the first line of effort.

Why have these efforts not been undertaken more widely? Partly because there are so many known failures, in the sense that attempts to reform revenue collection systems are unpopular with politicians, logging companies, buyers and even the Treasury itself (for fear of the extra work), and the official and unofficial messages received by proponents have been unmistakable. As for alternative approaches to financing forest management, the problem seems to have been rather a lack of imagination.

Only a few forest services in the tropical moist forest region have economists, and they have generally been employed as accountants rather than planners. Planning and development sections are also rarities, and they too have tended to concentrate on short-range problems rather than matters which affect the implementation of long-term policy, and which might affect the immediate profits of the logging companies and cattle ranchers who finance political campaigns. To some extent this bias is of long standing and may be attributed to the pragmatic and practical natures of people who have been attracted to a career in forestry. What is now called policy analysis or policy research was in some countries considered a frivolous activity. This may be a consequence of the general political support for forestry which was given by colonial governments. That is, the need to justify the activities of the national forest service was not a serious concern.

However, this willingness to leave forest policies as generalized statements of government intentions has caused difficulties since 1950, as governments have moved towards national and regional plans with specific targets for five- and ten-year periods. The foresters have often found themselves untrained to compete with other branches of government in the struggle for targeted funds. The ensuing problems are at least partly attributable to the traditional ways of selecting and educating foresters, the topic of the next part of this chapter.

Educational factors

Personnel preferred for the colonial forest services had in general the same characteristics as those for the district administrators: a high degree of self-sufficiency, a willingness to live in isolated and often unhealthy areas for long periods, good organizing ability, honesty and a certain amount of idealism about tropical development. A liking for and ability in sport was thought to indicate a suitable frame of mind. Undoubtedly, these characteristics were important in their time. However, as the conditions of government changed in the tropical moist forest areas (which is not to say that rural conditions have changed), a greater degree of sophistication and specialization has been necessary.

Perhaps because of the small size of the colonial forest services, and the desperate shortage of territorial forest officers, the need for an intellectually strong headquarters staff was not appreciated as early as it should have been.

Although staff numbers were built up to some extent in the latter days of the colonial regimes, the continued emphasis on geographic coverage left the forest services unready to treat with larger and more specialized staff in other government agencies.

This author believes that the colonial forest services were also slow to appreciate the kind of national staff who should have been trained to replace them. The strong competition for scholarships to universities, mostly in temperate-zone countries, was almost invariably won by students born and bred in urban areas where the schools were better. Students with an interest in biology had the option of better-paid careers in agriculture or medicine. These students were not attracted to a profession associated with rural life, from which their parents might have only recently moved. Surviving texts of recruiting talks given to schools by colonial forest officers emphasize the joys of safari and communing with nature, and could be almost guaranteed to put off the brighter students.

Consequently those who picked up the scholarships reserved for potential foresters tended to be those who had failed to secure grants for more prestigious occupations. The students were not especially interested in forestry as such and often made a poor impression on the sub-professional staff when they went into the field on their return from university training. A common reaction by colonial forest services was then to press for the upgrading of the sub-professional staff (who shared the "safari" outlook) to the professional grades, rather than accept more of the unsuitable recent graduates.

However, this was no real solution, since the up-ranked former sub-professionals were even less able to make coherent statements of the complex case for forestry in the new atmosphere of national development plans.

Naturally there have been notable exceptions to these general statements. There are sub-professionals who have proved to be perfectly competent in professional grade work, and there were early national graduates who joined forestry because they wanted a rurally oriented career. Nevertheless, there has been and continues to be a marked lack of staff who can treat with development planners and treasury officials on equal intellectual terms. It is incontestable that this lack continues to depress the role of forestry in developing countries.

It is only fair to say that the statements in the last four paragraphs do not apply to some countries in South and South East

Asia. For example, the Indian Forest Service was the third most prestigious branch of the civil service and attracted recruits of high calibre. Furthermore, the competitive examination system for the Indian civil services gave preference to those with a high degree of mathematical ability.

However, so far only a few of the developing countries have been able to overcome this problem. Both Argentina and Brazil in the late 1960s and early 1970s had large-scale twinning arrangements between their major universities and counterparts in the United States. The sheer number of students who received part or all of their tertiary education in the USA allowed market forces to operate after only a few years, so that forestry is now a prestigious (and in the private sector, well-paid) profession, capable of attracting recruits of good intellectual ability. These successes were made possible because of the firm political commitment of the respective governments to engage in large-scale training, plus the ability of some of the huge US universities to absorb considerable numbers of students who might have had difficulty in fulfilling the entrance requirements of European universities.

More generally, however, the training conditions were and are much less favourable. The developing countries either lack forestry faculties or these tend to have very traditional courses, heavily loaded with basic science lectures and grossly neglecting the practicalities of land-use management and development planning. Real multidisciplinary training is usually absent or, if taught, is not reinforced with field studies and practice.

The temperate countries have tended to run down their general courses which were devised for colonial forest officers, on the grounds that the developing countries now have sufficient undergraduate facilities. The increasing number of North American universities which are establishing courses in tropical forestry suggests that the European ex-colonial powers have been over-sanguine.

Research information

The last aspect touched upon in the diagnosis section of this chapter concerns the availability of information to guide land-use management planning in general and forest management in particular.

There are numerous guides to the procedures for land capability classification, as well as ecological guidelines for tropical land development. The user of these guides needs information about the productive capacity of the land under different usages. This information is very largely lacking, as regards forest productivity. Although there has been considerable progress in the development of efficient methods for the static inventory of forest, and in the making of inventories for the planning of concessions and logging operations, the availability of data from dynamic inventories is very limited. This is true in spite of the establishment of many thousands of permanent sample plots in tropical moist forests.

It must be remembered that, when the tropical forest services were started, information on the productivity of the forests was not of high priority. Emphasis was on the selection and establishment of the permanent forest estate. Protective forest was gazetted where current land use was damaging its own or the downstream productive capacity of the soil. Productive forest was gazetted in areas where there was little or no competition from other land users, or where the putative users or claimants were persuaded that forestry was more appropriate. There does not seem to have been any call, or attempt, to quantify the long-run productivity under different types of land use.

However, this situation changed towards the end of the colonial period and in the early years after independence. The change found the foresters ill prepared. There was a rapid expansion of the area under industrial tree crops as well as a considerable enthusiasm for settlement/colonization schemes for the landless: those who have never had land, and those who have lost it through political action against them. In many cases, these schemes were implemented at the cost of the forest, since excision from the gazetted permanent forest estate required no compensation and incurred no adverse political consequences. On the contrary, there was a bonanza for the forest industries, with huge quantities of top-quality logs available at low cost from the forest clearances. Not rarely, the pace of logging far outstripped the conversion to planned agriculture and even now, in some countries, there is devastated forest which was scheduled for conversion long ago, logged in a careless manner and then simply abandoned.

How did the national forest services come to lose these portions of the permanent forest estate? In the first place, they were usually gazetted long before there was any formal procedure of land capability classification (LCC). When the hunt began for legally unencumbered land for agricultural development, using

LCC surveys, the forest was often treated as a residual use for the land. It was common for the forest to be classified in terms of the static inventory data, that is what was easily saleable from a few species in large log sizes at that particular time. Given the low intensity of commercial exploitation in most of the tropical moist forest when the Far Eastern and European markets were booming, this single classification inevitably indicated that the forest was unproductive compared with any agricultural use of the land.

The foresters usually had little or no information to indicate the productivity of the land in terms of the multiple use of the forest. Although such use had been clearly recognized during the colonial period, and was the theme of the World Forestry Congress in Seattle in 1960, the foresters had been slow to quantify the multiple benefits of sustained-yield forestry.

The sudden demand for large quantities of economic data coincided in several countries with the period of independence and the loss of long-serving and qualified staff of considerable experience. Even though the developing countries have been expanding the numerical strength of their forest services, they often lack the experience which the older organizations had, with their balanced distribution of age classes in the civil service. The new, or newly independent, forest services necessarily promoted staff with great rapidity. As a consequence, senior officers may now have spent very little time in the forest and lack a feel for their work. These same staff may also have missed the opportunity to improve, under experienced supervision, their real capacity to handle the many and complex tasks of a senior forester. In other words, restoring a balance to staff structure and experience is a generational matter. It was bad luck that the small forest services were almost inevitably caught at a disadvantage when loss of senior staff coincided with a great demand for just that experience.

Even information on production for saw-timber alone was scanty and incomplete. Yet there were many hundreds of permanent sample plots from which such information might have been derived.

However, there were many technical deficiencies in these plots. It must be recalled that colonial forestry started in those tropical and subtropical areas with a monsoonal climate where a significant proportion of the commercially valuable species form annual growth rings. It was, at least in principle, easy to apply European methods for estimating age and growth from stem analysis. The situation is different in the moist tropics, where no commercially important species form annual rings in a reliable manner. The age of a tree cannot be determined from an examination of its cross-section. The need for repeated measurements on permanent sample plots was recognized at an early date but the complexity of handling and interpreting the data was not. The extreme variability of growth of trees in the tropical moist forest was unappreciated for many years and the mathematical techniques known to tropical foresters were completely inadequate for coping with the situation. Attempts to apply stand projection techniques have been continued by staff on FAO projects within the last decade and are no more successful than earlier ones because of the need to apply so many subjective "fudge factors".

Only in the Queensland rain forests of Australia has there been a recent vigorous attempt to catch up on modern techniques for growth analysis and prediction, using methods developed in the polyspecific multi-aged forests of the north-central states of the USA. Unfortunately, in other tropical countries, the national forest services have failed to produce estimates for forest productivity which will withstand the critical scrutiny of other competitors for land.

Clearly, if the forest services are unable to produce technically reliable justification for retaining land in the permanent forest estate, they are liable to be swept aside by competing forces. It does not matter that these forces may also have weak technical grounds for proposing a change of land use; the point is that an inadequate technical defence, added to the lack of the legal encumbrances of private ownership, make the tropical moist forest a very desirable location for any project which needs a large area of land.

Thus it is not surprising that many tropical forest services have lost a considerable part of their former forest estate. Their arguments based on the long-term public good compared with short-term private gain have been too general and philosophical for the relatively unstable governments of the tropical moist forest regions. These governments often feel the need for rapid action to defuse critical situations such as demographic pressure on land or the need to raise hard currency through the export of agricultural products. Their planners have not felt able to wait for foresters to produce more quantitative arguments for the retention of forests.

In the opinion of this author, the lack of good data on productivity and the multiple benefits of tropical moist forest is the principal cause of many continuing problems for foresters. Without such fundamental data the forest services cannot supply to planners and treasury departments any information about the costs and benefits of different actions. The sheer length of time needed to acquire comprehensive and reliable data clearly puts foresters at a disadvantage compared with agriculturists, but the problems are not technically insuperable. Foresters should be able to explain to the planners and economists why the data are not available, what resources are necessary to make them available, and when the results can be expected.

A parallel situation has prevailed on the utilization side of forestry. Prolonged efforts have been made to determine the anatomical, physical and mechanical characteristics of a great number of tropical timbers. However, the results are rarely presented in a form which is meaningful to end users, so the research has had little direct effect on the intensity of utilization. It should also be understood that much more effort has been devoted to research on the biological, "growing" side of forestry and much less to utilization of forest products. With hindsight, and in the light of the beneficial effects of demand for knowledge from an informed and responsible forest products industry, this emphasis should have been inverted.

The lack of usable information from forestry research, on both the growing and the utilization sides, seems to have arisen because research has often been treated as a scarcely necessary adjunct of the forest administration, a place where (in some cases explicitly) foresters who were poor at administration were shunted to be out of the way. There seems to be more than a grain of truth in the common view of foresters as people somewhat isolated from the mainstream of national development with not much to offer professionally; the reasons for this view have been mentioned above.

It is clear that many of the problems have arisen because the number of trained foresters is far below that required to manage the forests effectively. The problems become more acute as demographic pressures rise and governments seek land for the increased population, as well as resources with which to generate national wealth. The second part of this chapter offers some suggestions for improvements which bear on the management of the tropical moist forest.

PRESCRIPTIONS FOR IMPROVEMENT

Education and recruitment

That the number of foresters is low compared with the tasks they face has been recognized for many years. National foresters have not been struggling on their own, however. It is appropriate to mention here the many FAO projects which have been concerned with institutional strengthening to increase the number of nationals trained in forestry. Unfortunately, but perhaps inevitably, the counterparts on these projects have often spent an appreciable portion of the life of a project away on academic studies in third countries and have not benefited from on-the-job training. Given the small numbers of national staff available for attachment to an FAO project, the career advancement possible through the acquisition of an academic qualification has meant a general preference for overseas scholarships rather than on-the-job training.

There should not be much discrepancy between the two forms of training. However, the third-country institutions, mostly in the temperate zone, have rarely been able to tailor their courses to cover the specific needs of developing countries. Nor have they been able to cope with the different style of teaching to which the tropical students have generally been accustomed. Students who have spent all their school years, and perhaps taken their first degree studies, in huge classes being taught by repetition are unlikely to be able to cope easily with courses which place much emphasis on independent study and problem-solving. Such students need special assistance and supervision in order to develop their innate talents. The donor agencies which fund the attendance of tropical foresters at such institutions should exert more pressure to ensure that the large fees cover the cost of intensive supervision and intensive work with realistic case studies. This emphasis could be at the expense of covering a wide range of subjects; it is better for the students to learn thoroughly the context and application of a few basic techniques, plus the mental approach to problem-solving, than to have a superficial knowledge of a wide range of techniques.

The long and comprehensive forestry courses typical of continental European universities are simply not a viable option for small or developing countries. They cannot afford to have staff positions vacant for long periods. Equally, the financial cost of supporting students for four or five years is unattractive both to national treasuries and to donor agencies.

It used to be the pride of forestry training institutions that they turned out "jacks-of-all-trades". This is no longer a realistic aim but forestry academics and national forest services have been slow to appreciate the need for specialists. A forestry course which teaches house building, the alignment of roads and the measurement of trees is unlikely to have time to teach modern economics and the theories of social development. Attempts to include more and more courses within a curriculum in the 1960s were doomed. They failed to analyse the different needs of a developing national forest service compared with a stable service in a developed country, or to consider the optimum way in which these needs might be met. The problem was less that of overloading the timetable than of failing to see that the forestry profession needed to attract staff with a greater range of backgrounds.

There is no obligation to staff a forest service exclusively with those who have been through a forestry course. Undoubtedly there are difficulties of communication when staff with a mathematical or socioeconomic background work together with those from a biological basis. Shutting out the non-foresters has been disastrous for national forest services, because they have then failed to work meaningfully or harmoniously with other entities of government.

This is not to say that there is no longer a need for the traditional forestry training, far from it. The need rather is to make that training more effective. The absence of management studies, or an obligatory working plan exercise, from the curricula of most Latin American forestry schools is both symptom and cause of the failure on the part of many of these services to perform any meaningful function. Another much-needed change in the orientation of the traditional courses is towards a view of forestry as just one aspect of land use. An understanding of the advantages and disadvantages of different systems of land use would go far to diminish the mutual suspicions between foresters and farmers.

Policy development

Having ensured that a sufficient variety of training backgrounds is available to a national forest service (not necessarily all employed full-time), the leadership must ensure that the staff work within a relevant and meaningful policy. As indicated earlier in this chapter, qualitative statements on the benefits of forestry carry no weight with hard-pressed governments. The general statements have to be converted to action plans within the national planning and budgeting framework. The difficulty experienced by foresters in carrying out this task has also been widely recognized. The lack of data on which to base plans, due to failure to give sufficient emphasis to high-quality research, is reparable in the long term. In the short term, the FAO Tropical Forestry Action Plan (TFAP) offers a policy context covering the five major areas of concern to tropical countries.

Within the global framework of the TFAP, an increasing number of tropical countries have embarked on reviews of their forestry sector and its place in the national development plan and economy. Although some countries have carried out this review using entirely national resources, most have used some combination of national staff and consultants funded by donor agencies. The fairly lengthy review process is intended to provide the basis for a national forestry development plan, set in context and covering the period up to at least the year 2000. It is clear that some forest services have not been able to engage the attention of their ministers or the national planning and financial departments of government. The preparation of the national forestry plan in language which is understood by planners and treasuries should help to enhance the status and effectiveness of a national forest service. It is important for a country to realize that the preparation of a plan is only the first step in a prolonged effort, and that this effort must come largely from national staff. For countries where forestry at present has a low status, in spite of series of externally supported projects, the uphill path appears to be endless. In these cases, this author believes that the selection and basic education of the foresters is at fault. There may also be serious problems with career structure, or lack of it, and with the inability of many donor agencies to provide the sustained support for a long period which is necessary to build a self-sufficient national forest service in a small or poor country. The TFAP planning process provides a now generally understood framework which should allow longer-term views to be developed either by national foresters or by consultants.

Institutional development

The improved national forest services, with a more widely based staff, need to ensure that the multidisciplinary nature of forestry is understood inside and outside the service. A willingness to engage in discussion with other government agencies and non-government pressure groups would help to defuse many present tensions. This of course means that the staff must be sure of the technical basis of their own arguments and is a further ground for upgrading the calibre of the profession. At the same time, the service may have great difficulty in reconciling the need for a wide geographic coverage with the need for establishing a critical mass of brainpower at the centre to provide decision-makers with the necessary information. The civil service structures of many countries are notably inflexible and unrelated to changing needs.

A partial solution may be to farm out some responsibilities to agencies which are outside the civil service regulations. For example, there has been a tendency to separate research from the forest service, and bring it under a more or less independent institute, perhaps (and desirably) combining it with other aspects of land-use research. Another measure is to separate more clearly the regulatory and normative aspects of national forestry from field management, although this can be disastrous if the two or more agencies do not have clearly defined and complementary mandates, plus the functional ability to execute them. There is rarely any legal impediment to a forest service contracting out silvicultural operations to private companies, under forest service supervision, although the experience of the Philippines is that this is more successful with integrated forest industry companies than with smaller concerns. The major point is the need to consider a wider range of options than foresters have traditionally invoked.

Continuing the same theme, there is in some countries considerable enthusiasm for devolving managerial responsibilities to the local communities. In order for this to be effective, the forest service needs to have prepared demonstrations of suitable techniques, plus all the apparatus of an extension service. There is a surprising unwillingness to make use of agricultural extension services to promote good land use, with forestry as one component. The best demonstrations are often parts of formal experiments. It is lamentable that some TFAP exercises have recommended the creation of new extension agencies within national forest services rather than the reinforcement of existing agricultural capability.

Forestry research

Three aspects are emphasized here. First, in spite of the notable tradition of taxonomy and dendrology for tropical moist forests which has been established by several forest services, the assignment of basic studies to the increasing number of national universities should be considered. At the time when the herbaria, xylaria, insect and pathogen collections were established, the developing countries mainly lacked suitable universities, but this is no longer the case. Closer collaboration with the forest services might help the universities to provide them with more relevant education and better support, as well as relieving the government agencies of work which is difficult to reconcile with a modern system of administration. However, the proliferation of agencies, often because of political and personal differences, has been a source of extreme inefficiency in some developing countries. All that is suggested here is the need for an appraisal of the current methods by which basic research is undertaken.

In those countries which have a formal programme of forest research, visitors are often surprised by the large number of registered studies in progress or proposed, compared with the paucity of relevant and usable results. Techniques of research management are generally very little used; this is all the more serious because of the relative youth and inexperience of the operational staff. There needs to be better understanding of what research can and cannot achieve, and what resources (including time) are required to produce usable results. It is rare for undergraduate courses to cover research techniques, yet many require a research study as part of a degree course. This is a mistake, if the study is instead of a management plan, since far more graduates will go into management than into research.

Research staff need training in research techniques which are not restricted to statistics and computing. Particular emphasis needs to be given to the definition of variables to be studied, the collection and validation of data, the systems for handling and analysing data, and the reliability of the expected results. These matters should be laid out in formal study plans, to be reviewed by research directors according to a formal and scientifically defensible procedure. At the same time, research administrators need training in the setting of policy and the management of diverse operations.

As regards management (and, by implication, silviculture), the greatest need is an appreciation that forestry is highly location-specific. However, it is clearly impossible to test all possible management regimes on all sites. Therefore research needs to be planned in terms of its contribution to the development of models from which management regimes can be extrapolated, and which policy and planning staff may use to test the effects of (for example) changes in royalty rates or cutting regimes. The presently widespread proliferation of unco-ordinated silvicultural studies is entirely contrary to this approach and is a thorough waste of scarce resources. This is especially true for tropical moist forest, where research is often physically arduous, research sites are often remote, and the analysis and interpretation of experiments are complex.

Much more attention needs to be given to the management of data which have been so expensively collected. It is not at all uncommon for studies to be repeated at short intervals because the data from previous attempts have been lost. The increasing availability of low-cost computing facilities and data-banking systems greatly facilitates the mechanics of this process. The failure to make use of these advances is often due to the inexperience of the research administrators. The development of management models is of course conditional upon the creation and maintenance of data banks of rigorously validated data, mostly from repeatedly measured yield plots and experiments.

In the planning of research, the dissemination of the results to line managers and extension services is often forgotten. As previously mentioned, the most convincing demonstration of a technique may well be found in a well-planned and well-executed formal experiment, since there will be nearby examples of unsuccessful techniques (experimental treatments). Where there is a choice, the experimenter should choose formal designs which are easily appreciated by visitors without a background in research. For example, systematic designs for dose-rate experiments are visually attractive and effectively self-explanatory; "dose-rates" include inter-tree spacings, light intensities, fertilizer rates and some agroforestry combinations.

Gaps in information for management of TMF

It is perfectly possible to practise a conservative management of the tropical moist forest without having a detailed silvicultural knowledge of the behaviour of the component species. Such management, based on direct observation by staff with a good biological education and an "eye for trees", was practised in India and Burma from the nineteenth century and was applied in many other countries in Africa and Asia, and to a small degree in the Caribbean. With the exception of Surinam it was scarcely tried in Central and South America.

When the desired output is large logs of a small number of species, the yield per hectare from polyspecific forest is of course low and the cutting cycle is long. If the species are sufficiently fast growing and easy to raise in nurseries, the productivity can be raised by enrichment planting and the increased yield per hectare may justify the increased cost. If the species are amenable to growth in open plantations, the yields may be yet further increased. However, the growth rate of individual trees may be little different from one system to another: the yield difference comes from the number of harvestable trees per hectare. The majority of presently commercial species from the tropical moist forest have large-diameter crowns at maturity, and there would be little advantage in establishing plantations at the necessarily wide spacings, even if the finance was unlimited.

The swing away from poly-cyclic to mono-cyclic management regimes, from the late 1940s to the late 1960s, was due largely to a realization that repeated cutting would be detrimental to the forest in the absence of very intensive control of exploitation damage. Logging companies have often prevented national forest services from including even minimal damage-limitation clauses in logging licences. The practical solution was to stimulate a single heavy felling and then close the forest while the next generation grew up from seedlings. This system relied on the presence of a sufficient number of seedlings of desirable species surviving the logging operation and producing a commercially adequate stocking for the second and subsequent cycles. In order to provide more uniform conditions for the second generation, those areas not opened up by the logging were often given a climber-cutting and girdling treatment to remove large trees of undesirable species. From the very small number of experiments undertaken on established crops, there was some indication that a pre-commercial liberation thinning part of the way through the growth cycle would prove commercially rewarding at the end of the cycle.

Since then, some countries have felt that the market acceptability of species is changing so quickly that the trees remaining after the first cutting, instead of being poisoned, are left to enhance the yield, or bring forward the time of the second cutting. This feeling implicitly accepts a more heterogeneous yield at the second cutting in terms of the species composition, and also a probable delay in the maturation period of the desirable species since the available growing space will be partly occupied by presently undesirable trees. It would be possible to test the potential of this management scheme by studying the residual stand structure after logging and applying known rates of growth and mortality. No such study appears to have been carried out, but the argument has been used to stop almost all silvicultural work in Sabah.

Similar simulation studies could also be used to test the "selection management system" (SMS, a silvicultural rather than a management system) now supposed to be in force in Peninsular Malaysia. Unlike the preceding Malayan Uniform System, the SMS relies on the survival of a large proportion of the trees left after the first logging and their ability to form a sufficient commercial crop for a second cutting at about half the rotation age. The Forest Department in Peninsular Malaysia is still codifying the field rules for the application of the SMS but the idea appears to be based on the Philippine system which was developed in commercially far richer forest than the Malayan hill dipterocarp areas. Simulation studies can be made in any situation where there are large sets and long runs of reasonably reliable data. For example, they could be made in Ghana to demonstrate that the "selection system" developed in that country in the 1950s was based on arithmetically inadequate calculations involving many arbitrary groupings and averagings of data (from which almost any results could be obtained), and lacking any recognizable validation or checks against independent sets of data. Other large data sets, such as those held by the major logging companies in the Philippines, could be worked or reworked with modern computer-assisted techniques to study the validity of the silvicultural and management systems which are supposedly in force. Desk studies, although complex and tedious, are much cheaper and faster than new field experiments. The tardiness of forest services in starting this work appears to be due to lack of suitably qualified staff, another example of the reluctance to bring in other disciplines even when the need is obvious.

Without doubt there is a paucity of well-designed and properly executed silvicultural experimentation in the tropical moist forests and the lack of reliable data has greatly hampered the forest services from developing management schemes. However, it is also true that a good deal of experimental data has been collected in recent years and has either not been analysed correctly or has not been analysed at all. This is due to the lack of sufficient numerate staff. The design of silvicultural experiments is not especially difficult. The handling of the inevitably voluminous data has become much easier with reliable commercial computing systems and programs. As long as the implicit objective is the production of large-sized logs, the time needed to obtain sufficient data for usable results is bound to be lengthy compared with experiments on short-rotation pulpwood and firewood forests. Forest departments need to explore intermediate analyses, probably using multi-variate methods. However, these will require more intensive assessments than have been customary.

Necessary work on crop development in large-scale silvicultural experiments needs to be supported by studies on the ecology of component species. While it is possible to manage the forest conservatively on the basis of general biological knowledge, more intensive yet sustainable systems will be possible only as more information is acquired about the quantitative responses to treatment from formal experiments. The knowledge required is no different from that for plantation species. However, if there are good multipurpose objectives for the management of tropical moist forest as a polyspecific multi-aged crop, then the type of experimentation will be somewhat different because of the complex ways in which inter- and intra-specific competition may operate.

The current enthusiasm of biologists and ecologists for "gap" studies (natural or artificial holes in the forest canopy) appears to reflect the timescale of research grants rather than a profound analysis of the critical missing segments in the data. These deficiencies are in the post-establishment phase of the second cutting cycle. The required data are on the growth and survival of the different species in relation to the characteristics of the individual stems and to both local and area competition. The characteristics of the stems include crown position

Table 6.1: The decision chart

1 national forest policy
2 legislative framework
3 land tenure
4 specific objectives of management for the forest in question
5 static inventory

        6a properties
           and uses of
           desirable
           species

7 dynamics of desirable species
        9 short-term studies on dynamics:
           – ecological tolerances
           – responses to silvicultural
             treatment
           – logging damage to trees and
             soil (need formal experiments
             to aid the development of
             response surfaces)

        6b properties of
           juvenile wood
           of crop tree
           species

8 routine
  diagnostic
  sampling
     (DS)

        10 long-term studies on dynamics,
           e.g.:
           – pollinators/dispersers/
             predators, their interactions
             with crop tree species and
             each other
           – regeneration banks (seed,
             seedlings, saplings)
           – growth and yield models,
             based on yield plots,
             inventories and DS

        6c possibilities
           of secondary
           processing

in relation to the local canopy, the crown and stem form, and the types and extent of damage sustained during the initial logging.

There are various reasons for the paucity of the data from plots of ten years and older: in South East Asia, many were destroyed during the Second World War; in both Africa and Asia, many were lost in the upheavals following decolonization; the conversion of much of the regenerated lowland dipterocarp forest in Peninsular Malaysia to oil palm plantations removed both the yield plots and their direct relevance to the remaining forest estate which is concentrated in the silviculturally more difficult hill dipterocarp zone. A further major cause of loss has been the very success of some experiments, the plots being felled deliberately by pirate logging crews because of their clearly superior growth. One cannot overestimate the importance of protecting the plots since they are critical for the development of models from which sound management regimes will be derived.

Because the conditions for forest management are so site-specific, as mentioned before, it is not helpful to make specific recommendations as regards particular countries; these can perhaps be derived from the various country reports. However, most countries with tropical moist forest lack the studies which have been outlined above and will not make much progress until they are established and producing valid data.

Biologists and ecologists tend to lay heavy emphasis on the shortage of usable information. The position papers for the UNESCO-IVIC (Instituto Venezelano de Investigaciónes Cientificas) meeting in Venezuela in 1986 indicate that there is almost everything still to learn. In order to counterbalance that depressing catalogue, the last section of this chapter suggests what are the critical aspects of silvicultural knowledge which must be filled for the development of sustainable systems.

Critical knowledge for sustainable management

Table 6.1 is a highly simplified chart of what a manager of tropical moist forest needs to know, in approximately chronological sequence. The numbers on the chart refer to the numbered text paragraphs that follow. (For simplicity the chart is confined to the IIED/ITTO primary management objective of the production of valuable timber on a large scale (tens of thousands of hectares) to feed a capital-intensive forest industry. The sheer size of areas under the control of a tropical silviculturist or a TMF manager forces them to accept a high level of heterogeneity in their forests which is augmented by the effects of logging operations. This difference in scale surely accounts for much of the difference in approach to problems between the TMF manager and the ecologist. The chart is further limited to the permanent forest estate, that is the forest which is reserved legally for the supply of forest products to fulfil national domestic (and perhaps export) requirements, in accordance with the national forest policy.)

1. National forest policy: dictates the broad outlines of work for the national forest service.

2. Legislative framework: places the forest service and forest operations in their legal context. Unfortunately the forest law is too often set aside by short-term considerations of political and personal pecuniary advantage (i.e. bribery and corruption).

3. Land tenure: the forest service may find it hard to determine the true traditional right-holders or landowners (especially in the South West Pacific), or there may be no simple answer, or there may be political interference with the operation of the statute law on land tenure.

(Note that difficulties caused to the forest manager on these three points are often due to the deliberate setting aside of the law by politicians.)

4. Specific objectives of management: these should take into account the demands on the forest, and the commitments to supply local consumers as well as large industries. The objectives are not set on a once-for-always basis but should be kept under review. Formal forestry working plans usually require a quinquennial or decennial revision. There should be feedback from the succeeding steps in the decision chart to ensure that objectives are adjusted to cope with long-term market changes and with improvements to the knowledge base for silviculture and management. The objectives may imply large-scale and year-round operations, or small-scale and perhaps seasonal operations; or a combination. Multiple phases of operations allow multiple benefits to be obtained. For example, a large forest industry might remove the big logs with heavy machinery and a second-stage licence might permit local people to remove residues for firewood and to collect minor forest products.

5. Static inventory: this is possibly the field in which scientific knowledge has had the greatest impact, since electronic computers took the drudgery out of sample calculations and data sorting and tabulating. Inventories are now multipurpose, to suit multiple objectives, but just as there has to be a primary objective so there must be a primary suite of variables to be estimated, which determines the sampling scheme. Sub-sampling examines the regeneration banks; nowadays that would include the seed bank in the soil. It is pleasant to record that forest-service silviculturists were studying the soil seed banks before university ecologists became interested in the subject.

6. The properties and uses of mature trees of species found by the inventory to be available in commercial quantities should be reviewed and researched. Note that in all, or almost all, tropical countries the number of species tested and found to be industrially suitable by forest products laboratories exceeds the number of species actually marketed from TMF. The difference is partly a reflection of the strongly conservative marketing of the timber trade and partly an indication of the pressure which the forest industries business can bring to bear on the relevant minister. Subsequent studies examine the properties of juvenile wood of the second-crop species, to see if an early harvest would provide technically adequate timber; more advanced studies look into the possibilities of mixing species in a single processed product such as chipboard or paper, as well as methods of adding value by secondary processing (such as overlaying printed films and plastics on to plywood, or selling furniture made from mixed species instead of only fine timbers).

7. Dynamic inventory: study should concentrate on the population dynamics of the desirable species. These are defined primarily on the basis of timber properties and secondarily on observed ecology (growth habit, position in relation to the canopy in appropriate seral phases after logging, growth rate).

8. Routine diagnostic sampling: DS is a generalized and improved form of the various linear sampling methods developed in Malaya, Nigeria, Sabah, Sarawak and Uganda. As the name implies it is used to determine the appropriate type of silvicultural treatment (if any). There is now more emphasis on early identification of potential final-crop trees ("leading desirables" – LDs) and a concentration on their liberation from competition. In many tropical countries there is a more or less explicit land policy of "use it or lose it", so there is

strong pressure for a forest service to re-establish its claim to manage the forest after a logging operation. An early silvicultural treatment, not necessarily poison girdling of undesirable species, is thus often politically desirable. However, the forest service needs to establish the scientific worth of the treatments applied after best-guess interpretations of the DS.

9. Short-term studies on dynamics: as in DS the emphasis is on desirable species, considered together as a crop, rather than on the individual species. Three classes of studies are especially indicated by forest managers: ecological tolerances, determined by sample surveys in space and time as well as by formal experiments; the response to silvicultural treatments, including response to the major intervention which is usually the logging operation itself. Failure to quantify the pre- and post-logging states of the forest before the application of experimental silvicultural treatments has resulted in a regrettably large number of uninterpretable experiments; and the effect of actual and simulated logging damage to standing trees and to the soil.

The emphasis should be firmly on experiments rather than on observation. Hypothesis testing is rarely necessary; it is the quantification of the response which is needed. Results of the trial should be incorporated into growth models (based mainly on yield plots, see step 10) whose sensitivity is often determined by the quality/quantity of data at the extremities of the site/silviculture/growth response surface.

10. Long-term studies on dynamics: these are intended to provide the bulk of the data needed for growth modelling and yield control systems. The main data source should be a well-stratified and regularly measured system of yield plots, supplemented by occasional inventories to improve spatial coverage, and making provision for incorporation of routine DS results. The high rate of turnover which is now known to apply to natural forest previously thought to be very stable applies also to LDs. This makes modelling difficult – if not impossible – f ,m data collected solely from LDs. It would seem that there is no satisfactory alternative to recording data from yield plots for all stems of desirable species over a defined minimum size. Much more effort needs to be invested in growth studies and modelling, since the failure to make secure predictions of future yield renders forest services vulnerable to arbitrary political decisions concerning logging operations. Three classes of studies

would be particularly helpful to forest managers: the critical conditions for pollinators/dispersers/predators of the crop tree species,
and their interactions with the trees, as well as with each other; the
dynamics of the regeneration banks (seed, seedlings, saplings/poles);
and the ecology of climbers (lianes, trepaderos), particularly those of
silvicultural importance such as *Merremia* in South East Asia and the
South West Pacific and *Acacia ataxacantha* in West Africa.

Some managers would add a requirement for studies on nutrient cycling. Since 4 cu. metres/ha/yr is about the best commercial
growth rate that could be obtained in lowland TMF, and since 10 cu.
metres/ha/yr seems to be about the rate above which some artificial
fertilization would be necessary to sustain yields, nutrient studies are
not high in this author's priorities.

Routine diagnostic samplings may indicate that some areas of forest
have less than the minimum number of potential crop trees in the
regeneration. Depending on how the critical stocking levels are set in
the DS interpretation instructions, the prescription may be to enrich
the forest with line plantings or to replace it entirely with artificial
plantations. The latter may also be prescribed if demographic pressure or market demand increases so much that the natural forest
must give way to a more directly productive form of land use. Such
conversion is not necessarily an indication of managerial failure; the
forest may have been damaged by natural or human forces before it
passed to the control of the forest service, or the demand may exceed
the biological capacity of the natural forest to produce the desired
materials.

Departures from the managerial process outlined above are caused
by social, economic and political pressures. Plantation forestry has
suffered as much as TMF from budget failures and land tenure problems; maybe more, because plantation forestry is the management of
intentionally unstable systems and requires timely interventions to
prevent their collapse.

In summary, although ecological knowledge implicitly underpins
forest management, in an explicit form it is only one of a number of
factors influencing TMF management. Ecologists might have more
influence if they interpreted their research in terms of potential
impact on management, while managers should articulate their
research needs more clearly and phase them into grant-sized projects
as understood by academic ecologists.

CONCLUSION

Little of the foregoing is covered explicitly in the recent literature, for reasons given on pp.154-5. However, unless and until these problems are tackled in a professional way the management of TMF must remain conservative if it is to be sustainable in the long term. Sustainable increases in productivity in TMF depend, as in every other field, on valid research, maintained over the whole timber rotation. No agronomist is expected to predict maize yields from the first ten days of a 90-day crop, nor can foresters predict yields of 90-cm diameter trees after only 10 cm of growth. Much more could be done with existing information if the forest services made use of a wider range of backgrounds when recruiting their staff.

It is perhaps worth noting that points made in this summary may be found in some of the as-yet-unpublished reviews of national forestry sectors, prepared under the FAO Tropical Forestry Action Plan. The general principles of forest management have not changed since organized tropical forestry began over 130 years ago. Unfortunately, neither principles nor techniques are taught to the foresters in many countries with tropical moist forest. Foresters themselves can redeem this situation and should not look for scapegoats outside the profession. As pressures increase for more intensive use of the land, the foresters must become more skilled and more adaptable if they are to retain the forest on which their livelihoods depend. There is a great need for better operational direction and a more forward-looking structure for career development in all countries which aspire to retain their tropical moist forest.

NOTES

1. Only minor corrections and additions have been made to the text submitted as part of the IIED report to ITTO in 1988. These were made mainly at the instigation of J. Wyatt-Smith and H.C. Dawkins.
2. Meadows, D.H., *The Limits to Growth* (New York: Universe Books, 1972).
3. Dawkins, H.C., "The first century of tropical silviculture – successes forgotten and failures misunderstood", in Melanie J. McDermott (ed.), *The Future of the Tropical Rain Forest*, Proceedings of an International Conference, St Catherine's College Oxford, 27-8 June 1988 (Oxford: Oxford Forestry Institute, 1988).

7. Conclusions
Duncan Poore

In this chapter the results and conclusions of our study are presented, and are followed in the next by our recommendations. These were addressed in the first instance to the International Tropical Timber Council at its meeting in Yokohama in November 1988.[1]

THE FINDINGS OF THE STUDY

Introduction

At the end of the first chapter we set out the principal questions to which we wished to find answers. They were:

(a) Over what areas is natural forest managed at an operational scale for the sustainable production of timber?
(b) Where such management has been undertaken successfully, what are the conditions that have made this possible?
(c) Where such management has not proved possible or has been attempted but failed, what have been the constraints that have made it difficult or impossible to apply? (The conditions for success and the constraints tend to be mirror images of one another.)

It is evident that the answer to the first question – the amount of forest under sustainable management – depends entirely upon the criteria used in determining whether management is in fact sustainable. Definitions are therefore critical, and the reader is referred back to the discussion in Chapter 1.

A number of points were made there that are sufficiently important to bear repetition. For example, a distinction was made between the characteristic tools of management on the one hand and of silviculture on the other. It was pointed out that "management was often said to be uneconomic when it was meant that silvicultural treatments would be"; and that demonstrations and trials were often given the unjustified status of "management projects". In the analysis that follows we are, accordingly, concerned with *management for sustainable timber production*, not sustainable management for other purposes such as catchment protection or nature conservation, and *management at an operational scale* rather than demonstrations or trials.

We also suggested that there were various levels at which forest management could justly be termed "management for sustainable timber production". But we can only consider them as truly operational if a conscious decision has been made to manage the forest at that particular level, if this decision is being conscientiously applied, and if the results are being monitored. This implies (what is in fact the case) that there may be a number of areas which have not been brought within any formal management structure but are not deteriorating. The necessary adjustment can readily be made in such instances to bring them under formal sustainable management.

There are also some other questions which we need to make clear if our results are not to be misunderstood.

First, it is not yet possible to demonstrate conclusively that any natural tropical forest anywhere has been successfully managed for the sustainable production of timber. The reason for this is simple. This question cannot be answered with full rigour until a managed forest is in at least its third rotation, still retains the full forest structure, is fully stocked with commercial species which are growing well, and possesses adequate regeneration and an intact soil and ground flora. No tropical moist forest has been managed consistently for a sufficiently long period to fulfil all these conditions. But it is not particularly helpful to come to this conclusion and we adopt, therefore, a definition which is strict but not so stringent as to have no utility.

Second, one should appreciate that, whatever definition is adopted, the management status of the forest may be altered almost overnight by changes in the implementation of government policies; and this alteration may be in either direction. The best-managed forest may be arbitrarily allocated to another use – an agricultural development scheme perhaps, or even a plantation of fast-growing trees for pulpwood; or it may be decided to make a sudden and unjustifiable change in the length of the cutting cycle. At the stroke of a pen a forest which was managed sustainably becomes one which is no longer so managed. In contrast, a forest which is outside state control may be demarcated, provided with a management plan and brought under effective control. It then can be considered to have a reasonable chance of being managed for sustainable production in the future. To this extent any figures which we give represent the best judgement that can be made on the prospects of areas under different forms of management. Changes in the implementation of policies could readily lead to the figures being revised substantially either upwards or downwards.

For this reason it is vital that governments should fully understand the present situation; for management that is good could very easily be turned into management that is destructive and, on the other hand, the very substantial areas of forest that are now very nearly under a sustainable management system could very readily be turned into forests which are being managed for sustainable production in a satisfactory and convincing manner.

Accordingly, in this analysis we have adopted a strict definition of forest management: that it should be practised on an operational rather than an experimental scale and that it should include the essential tools of management, these being objectives, felling cycles, working plans, yield control and prediction, sample plots, protection, logging concessions, short-term forest licences, roads, boundaries, costings, annual records and the organization of silvicultural work. They must also meet wider political, social and economic criteria without which sustainability is probably unattainable.

Some of our critics consider that this definition is too strict; others think that it is too lenient. Being criticized from both sides makes us think that we may have got it about right. If we had used a more relaxed definition, it would have bred attitudes of dangerous complacency; what is needed is to generate a sense of great urgency but qualified optimism.

The results of our study, using this strict analysis, provide a picture that is not encouraging. But, before we give the figures for the different continents, we should point out that there is a brighter side to the picture as well. There are many areas in which certain elements of management are in place. There are many silvicultural trials which have produced encouraging results, and many signs that countries are taking some of the steps which are needed to establish sustainable management at an operational scale. Examples will be found in the preceding chapters and in the country reports. Others are mentioned in the extensive reviews prepared by FAO covering Asia and tropical Africa.[2]

A few, quoted below, illustrate how important elements of

sustainability can be maintained in forests managed at very different levels of intensity, although in none of these cases has sustainable management yet been convincingly demonstrated in practice.

Wait and see

This policy is being deliberately applied in the Tapajós National Forest in Brazil. The forest is reserved; knowledge is available upon which to manage it; but operational management has not begun because there is not yet a profitable enough market as an abundant supply of very cheap timber is taken from land being cleared for agriculture.

Log and leave

Forests treated in this way cover very large areas in all continents. This would be an acceptable management regime if it were the result of a deliberate management decision based on reliable information about potential future crops and if it were adequately controlled.[3]

But unfortunately this is seldom the case. Frequently there is no proper security of tenure, no determination of logging standards or of the period a forest should be left before relogging. Log-and-leave forests are very vulnerable to abuse. In many African forests logging is very light and the enhancement of growth or regeneration through logging is slight; if left without apparent attention from forest departments, they are a great temptation to colonists. In other areas logging may be too heavy or there may be strong inducements to enter the forest to relog for other species or stems of lower girth and quality. (Both increased timber prices and the entry of "lesser-known species" into the market could increase these inducements.)

Minimum intervention

Such a system was being deliberately practised in Queensland. The possibilities of both stand improvement and enrichment planting were rejected on economic grounds. The resulting lack of intervention has considerable environmental advantages in preserving biological diversity.

Much more forest may be managed in this way in the future, as there seems to be a prevailing trend towards poly-cyclic systems and a recognition that stand improvement has often led to the elimination of species that later become marketable. The environmental benefits may also become better appreciated.

Stand improvement and enrichment planting
There are many examples, but almost all have failed so far to proceed beyond the project scale.

How much forest is under management for sustainable timber production at an operational scale?

Latin America and the Caribbean
In Latin America and the Caribbean the total area being sustainably managed at an operational level is limited to 75,000 ha in Trinidad and Tobago of which 16,000 have been "declared as fully regenerated after logging". Synnott states (pp.75-6):

> From the viewpoint of professional forestry, this author has not identified any case of operational TMF management for sustainable timber production in any member country except Trinidad and Tobago. Even in Trinidad, management is not intensive, but it does qualify as sustainable, even though silvicultural treatments are rarely applied and working plan prescriptions are not followed in strictest detail. In other countries, in spite of striking advances during the past ten years or more, the following components are generally weak or lacking: advance planning of the location and intensity of the annual cut; supervision and control to ensure that the cutting conforms to the planning; and protection of the area to limit unplanned activities including settlement and uncontrolled logging.

Africa
The conclusion for Africa is similar: that there are no sustained-yield management systems that are being practised over large areas in the six countries studied. Management has been progressively abandoned, maybe with the partial exception of Ghana. A selection system similar to that in Peninsular Malaysia (see p.69) has been tested experimentally in the Côte d'Ivoire for eight years. Preliminary results are considered encouraging enough for the system to be applied to 10,000 ha of Yapo Forest.

Asia
It has not proved possible to give an estimate of the area of forest which is under genuine sustained-yield management in Asia. But Asia differs fundamentally from both Africa and America in that,

with the exception of Papua-New Guinea, all the forests under concession agreements within the region are, at least nominally, under management.

There is, however, a very great difference between theory and practice in many parts of the region. In Chapter 5, Burgess identifies the needs for planning and control as follows: (1) yield or coupe confined to prescribed limits; (2) fellings orderly and confined to coupe boundaries; (3) residual stand adequate; (4) residual stand protected; (5) silvicultural work carried out; (6) felling cycle followed, relogging not permitted between predetermined cycles; (7) permanent roads maintained and post-felling erosion controlled; (8) unworked forest protected; (9) working plans written and enforced. He goes on to remark: "The items on the above list where management tends to be inadequate are 4, 5, 6, 7, 8 and 9".

This difference between theory and practice is impossible to quantify without much more detailed inspection than was possible in the time available for study.

It is possible to say, however, that the most encouraging and complete theoretical system for operational management in the region is in place in Peninsular Malaysia where the so-called "selective management system"[4] provides for stand modelling based on adequate inventory and the data from permanent sample plots (PSP). Various kinds of silvicultural management can then be applied, all having the objective of producing a sustainable yield. The system has only recently come into full use and there is no certainty of its success into and after the first cutting cycle, but it contains many of the elements which are necessary for a flexible response to the prevailing condition of the forest stand. It is intended to apply this system to the total reserved production forest within the legally constituted Permanent Forest Estate of Peninsular Malaysia.

Burgess has identified (p.144) the following as "forests and operations [which] appeared to be reasonably successful as potential sustained-yield units": in Thailand, the Mae Poong forest; in Peninsular Malaysia, many candidates of which Jengai Forest Reserve, Trengganu, is one of the earliest examples; in Sabah, some of the operations of the Sabah Foundation; in the Philippines, the operations of the Nasipit Lumber Co.; in Indonesia, the operations of Padeco, Sungei Rawas, and the Musi River, Sumatra.

Queensland, Australia

In Queensland just over 160,000 ha, the whole area of forest scheduled for logging, was until recently under a management system for the sustainable production of timber which was soundly based on research results, could respond flexibly to the condition of the forest stand and was under strict control. In addition it was subject to stringent environmental guidelines and went very far towards ensuring that the disturbance to the whole range of species of forest-dwelling plants and animals was kept to the minimum. In spite of this, however, logging has now ceased because the Commonwealth government considered that it was inconsistent with the nomination of the Queensland rain forests as a potential World Heritage Site.

Overall picture

It is possible then to say that the area of tropical moist forest which is demonstrably under sustained-yield management for timber production in the producer countries of ITTO (with the exception of India) amounted, at the very most, to about 1 million ha until the decision was made about Queensland. It has now, on environmental grounds, been reduced by about one-fifth. This is out of an estimated total area of some 828 million ha of productive tropical forest remaining in 1985 in all the countries in which it occurs, not only those belonging to ITTO.[5]

Under these circumstances urgent action is required, not only to ensure the proper management of previously unlogged forest, but also to assess the status of logged forest and degraded forest lands and to plan remedial action to bring these under sustainable production as rapidly as possible. This is the remit of other pre-project studies commissioned by ITTO.

THE CONDITIONS FOR SUSTAINABLE PRODUCTION

Much of the time of the consultants and much of the attention of the round-table discussions held in each of the continents was devoted to identifying the conditions under which management for sustainable timber production had hitherto been successful or unsuccessful. Not surprisingly, these turned out to be mirror images of each other and so could be treated together. We then attempted to define the conditions which were necessary to give the greatest chance of success in the future and, as a move in the direction of setting priorities, to arrange these in some kind of hierarchy. They are: (a) long-term security; (b) operational control; (c) a suitable financial environment; and (d) adequate information.

Long-term security of operation

A permanent forest estate for timber production

It would seem a truism to state that management for sustainable production is only possible if the forest continues to exist; without forest there can be no production. But we found that the lack of any guarantee that forest would remain as forest was the overriding reason for failure in many instances; and that, moreover, the absence of any guarantee of security of tenure discouraged forest managers, whoever they might be, from investing time and money in future management.

Three very different examples may be quoted to illustrate the point.

In Peninsular Malaysia there was developed one of the potentially most effective silvicultural systems,[6] the Malayan Uniform System, to manage the very productive lowland dipterocarp forests of the peninsula. Early results with this system were most promising. But the forests to which it was applied were situated on soils which were eminently suitable for agricultural cash crops and, as a matter of government policy, appropriate stands were alienated after the first cutting cycle for conversion to plantations of rubber and oil palm.

In Ecuador a number of logging concessions, which were operating until 1970–80 in defined areas with defined yields, were cancelled because of the practical and political impossibility of protecting the concession areas from occupation and cultivation by colonists. As a result of this experience, the government has effectively abandoned the use of concession agreements as a tool of forest management.

In Queensland, Australia, a system of permanent forest reserves has been defined by legislation and the boundaries marked on the ground; this was part of a land-use policy which also allocated a substantial fraction of the forest area for nature conservation and for catchment protection. A proportion of the area so reserved was allocated for the harvesting of timber and a system of "minimum intervention" logging was devised which is the most complete example of sustainable management that these consultants have found anywhere in the tropics. Yet, as mentioned above, because it is now proposed that the whole of this area should become a World Heritage Site, timber harvesting has been suspended by the fiat of the Commonwealth government.[7]

Other examples could be quoted, but these will suffice to illustrate some general principles. All the other experience of the consultants is consistent with them.

In each of these three instances central government has not had a firm and consistent policy with regard to maintaining a permanent forest estate for timber production. This has meant that it has either changed its policy or has willed insufficient resources to defend its policy. In the case of Queensland a conflict between state and federal interests has also been involved.

In two of the cases cited, a sector of the population considered it more important to replace the use of the forest for timber production with some other use: in fact the policy of sustained-yield management failed to carry an influential section of the population along with it. In the case of Ecuador the forest was invaded by colonists who saw a greater advantage to themselves in converting it to farming land; in the case of Queensland the opposition to logging came from environmentalists.

Some general conclusions can be drawn from these experiences.

First, governments of the producer countries should consider it a matter of high priority and importance to set aside, as a part of an overall land use policy, a permanent forest estate (as has recently been done by Peninsular Malaysia) which it should be prepared to defend both practically and politically. It is, of course, impossible that governments should be able to identify in advance all claims that may arise in the future to alienate parts of this estate (especially where these may be quite unreasonable), but there are a number of general trends that cannot be ignored. Among them are the following.

As the population in the producer countries rises, more land will certainly be required for the growing of food. It is likely, therefore, to be a self-defeating policy to include within the permanent productive forest estate extensive areas that are highly suitable for the sustainable production of food. It is also a legitimate economic decision to decide to use land for the sustainable cultivation of non-food cash crops rather than timber as a part of a policy of economic "take off".[8] In an extreme form this may, after careful consideration of the implications, involve a deliberate decision by a particular country that it will not remain a timber exporter or even a timber producer.

There will be increasing pressure from an environmentally sensitive public in four directions: to ensure that national land-use plans make adequate provision for the protection of catchments; that the rights of groups of indigenous peoples are respected; to grant protection to an adequate variety of the ecosystems of the country and to its plant and animal species; and to organize management of the forest for timber production in a sustainable manner and in such a way that it does not cause environmental damage. It would be a mistake to assume that this pressure is mainly generated outside a country; experience shows that the growth of environmental awareness is an inevitable concomitant of development.

There are also likely to be claims for additional physical developments – roads, towns, reservoirs, etc. (In Peninsular Malaysia, now that a Permanent Forest Estate has been defined, developments such as these are the outstanding issue.) Because of the need for permanence in the forest estate, any claims of this sort should be subject to the very strictest scrutiny.

Because of factors such as these, it will be necessary to have politically cogent arguments to justify the size and location of the permanent productive forest estate against the claims of other land users – in particular claims for the growing of food and for environmental protection and on behalf of indigenous peoples. Without security sustainable production on forest land is unattainable.

Three parties need to be convinced: the economic and planning ministries of governments, that the goods and services to be derived from national forests and forest lands are of high importance for the future well-being of the country as a whole and of parts of the whole; local populations, that they can derive greater benefit from well-sited national forests than from alternative forms of land use; and environmentalists, that there is benefit in managing some forests for production and that this management can be carried out in a sustainable and environmentally acceptable way.

We fully agree with the conclusions of the HIID study[9] that governments consistently undervalue the timber resources of the forest and the environmental services of forest lands, and that they rate too highly the value of converting forests to other uses.

Security for the managers
So far it has been assumed that the forests will be owned by the government and will be managed by government officials or under their direction. In the three cases that have been mentioned so far

(Peninsular Malaysia, Queensland and Trinidad) this is indeed the case. In all countries examined in this study, with one exception, there are substantial areas of state forest or there is provision in government policy for there to be so. (The exception is Papua-New Guinea where the majority of the land is legally vested in local communities.) State ownership and management can evidently be satisfactory, but only if the state has the will and the capacity to exercise effective control.[10]

This, however, is not the only possible model and we have found in our study that there are promising signs that forests can be managed sustainably by other agents if the right conditions are provided; and, conversely, that security of tenure, though essential, is certainly not enough to ensure that government forests are well managed. These examples do not so far cover large areas but they have the elements of security, self-interest and incentive that seem to be necessary to ensure sustainability.

In Ecuador, for example, Synnott reports that

> a number of timber companies own forest land, notably
> the Empresa Durini. Many of these have logging op-
> erations which are more efficiently managed than those
> in unreserved forests with annual licences, and also
> industrial plantations and natural forests which are more
> intensively protected and managed than those on state
> land.[11]

Although he comments that these are not yet systematically managed for the sustained yield of timber, the potential and incentive to do so are evidently there.

In Brazil the company Florestas Rio Doce is working a number of separate small areas of tropical moist forest with an interest in regeneration and long-term production, with controlled cutting and monitoring of regrowth. And Jarí Cia Florestal Monte Dourado, a company with a high reputation for the technical quality of many of their operations, has recently begun to take a serious interest in systems of partial and selective exploitation in TMF and in monitoring their effects.

Secure ownership by forest management companies, under clearly defined government conditions that the forest must remain as "natural forest", might therefore provide a satisfactory alternative to state ownership and management. Certainly the converse seems frequently to be true: that the insecurity of present concession agreements, and the fact that these are almost invariably for periods much less than the rotation or even the cutting cycle of the forests that they cover, do not provide sufficient incentives to concessionaires to protect their forests from encroachment or to invest in sustainable management.

Another possible model is the leasing of the land to local communities for management under defined conditions. There are successful examples of this in Quintana Roo, Mexico, and it has been suggested for Indonesia.[12]

This is clearly the only model that could succeed in Papua-New Guinea, although there is little sign so far that satisfactory conditions have been created there to interest the local populations in keeping forest as forest, far less in managing it for timber production. The greatest potential for this kind of management that we have identified in this study are the "Areas de Manejo Integrado" in Honduras. "These consist of a kind of social forestry system, involving the planned and rational use of the forest by local communities using simple techniques, including timber production by pit-sawyers" (pp.107-8). So far these have been established only in pine forest. There are at present fifty, each of between 1,000 and 10,000 ha, and 450 are planned by 1990. The Honduras forestry agency, COHDEFOR, intends to extend this model into areas of TMF.

Some interesting proposals were made by the FAO expert at the Asian seminar for handling the management of forest where it seemed impossible to attain a sufficient degree of government control over logging and forest management. He suggested three possible lines of approach when present concession agreements lapsed. First was the disposal of forest lots by auction for periods of 50 or 70 years. The forests would be inspected every five years and, if the report was favourable, the period of lease could be extended by five years, thus maintaining a future prospect of 50 or 70 years. Purchase would be supported by the banks under a system comparable to property mortgages. A second possibility was the lease of blocks of forest to provincial or lower levels of government to be managed by the forest authority for the benefit of the public. A third possibility was the lease of blocks to local communities for local use.

All of these are possible models for sustainable management provided that tenure is secure and that the sale or lease is conditional upon the forest not being converted to other uses – that, in spite of ownership and management being devolved, they remain part of a secure permanent forest estate.

Operational control

If the forest estate is secure, the most important condition for sustainability is the ability to control subsequent operations. By far the most important of these is logging; it constitutes the most severe disturbance of the forest ecosystem and is a necessary part of timber harvesting. If logging can be appropriately controlled and the logged forest then closed until there is an adequate crop for the next harvest, there will certainly be a next crop eventually, even without subsequent silvicultural treatment. Any such treatment is mainly concerned with further influencing the composition and marketable volume of the next harvest, its quality and, perhaps above all, the time that must elapse before it is taken. These are essentially economic questions, balancing present known costs against future indeterminate gains. Moreover, it should be possible to influence the harvesting operation to accomplish many of the silvicultural objectives.

We therefore argue that, after the security of the forest, the most important condition for sustainable production is the control of the whole harvesting operation. If no forest exists, management is irrelevant; but, with a secure forest, the sustainability of production can technically be assured by the control of harvesting alone. Subsequent silvicultural treatment is adding gilt to the gingerbread – desirable, but not essential.

The significant conditions, therefore, are: clearly defined management objectives for the forest and a plan of management; standards and defined procedures; adequate control of harvesting based upon both these; and an assessment procedure to determine the initial average length of the cutting cycle and any subsequent refinements of it.

Control of harvesting is necessary for two separate reasons: the regulation of the future crop and the reduction of adverse environmental side-effects. The former is mainly concerned with the quality of residuals and advance growth, the presence or absence of regeneration and (possibly) the maintenance of site fertility; the latter with dangers of soil erosion and the possible reduction of the variety of species and of genotypes.

As far as the future crop is concerned, management should remain flexible; the behaviour of the forest and of the market can be predicted, but imperfectly. There are obvious merits in a flexible predictive model for management which operates through the setting of standards and through control – of such matters as girth limits for felling, number of residuals and composition of residual stand, regulations about leaving or removing damaged trees, species of lesser value or potential seed trees, differential stumpages and many other features. All of these can be manipulated through the control of harvesting.

Much less is known about the effect of harvesting and subsequent management on "biological diversity". But, as information becomes available, it can be added to any such dynamic model so that, in theory at least, environmental advantages can be set against additional costs or reduced harvestable yield. The same considerations apply to other, non-timber forest products.

The most evident environmental effect of logging is, however, erosion. This can be controlled by rigorous standards of design and control for roads and use of equipment, such as those in Queensland. Benefits, costs and practicability are, of course, factors in the setting of these standards; but in many cases better roading and machinery lead to the possibility of extraction at lower cost.

It should be emphasized that standards, control and prediction all depend absolutely on adequate information of the right amount and quality; and a system of control – whether operated by government, private owners or communities – that is reliable. It need hardly be said that the requirements for information and for control become the greater, the more sophisticated the management.

It is difficult to say whether lack of security or lack of control is the most common reason for the failure of sustainable production; but it is unfortunately true that lack of adequate control is almost universal outside – and even to some extent within – those rare examples quoted above where operational management is succeeding. It can take many forms: not allowing adequate recovery time after harvesting; not marking trees for felling or doing so negligently; undue logging damage; insufficient residuals; insufficient regeneration; relogging too early; illegal logging; concessions given to unsatisfactory operators or for the wrong reasons; failure to exact penalties for faulty logging practice; and many others.

As the vast majority of operations is ultimately controlled either by government departments or government agencies, the failure of control must ultimately be laid at the door of governments. The causes are staff shortages or low staff morale; poor conditions; insufficient equipment; inadequate field supervision; inadequate or inappropriate training; and insufficient research information. In an extreme form, these can mean that the lives of staff are endangered if they show themselves too conscientious.

All these are symptoms of the low ranking of forestry in the priorities of governments and also in the view of many influential citizens, all of whom who tend to view the forests as a resource to be mined, as constituting a residual use and as less valuable once exploited than for alternative uses. Many of the prescriptions in Chapter 6 are rightly concerned with these issues.

A suitable financial environment

The third condition for sustainable production is a market for the produce. The absence of a market is the main reason why large areas of forests in the Central African countries remain unexploited; and the low volume of timber removed from them is cited as one of the principal causes of the absence of adequate regeneration in them. As already mentioned, the management plan for the Tapajós forest in Brazil has been put on ice because there is not yet a sufficiently lucrative market for its timber. On the other hand, the flexible and successful management for stand improvement in Trinidad was largely made possible because a local market for charcoal was skilfully used by the forest managers to carry out silvicultural operations at low cost.

But the presence of a market by no means ensures that management will be sustainable. There are unfortunately too many examples to show that the opposite is usually the case; there is a critical difference between self-interest in exploitation and self-interest in sustained management – a difference that must be borne in mind by those designing forest policies. The latter must always ultimately depend upon the resolve of governments, which may be manifested by investment in their own forestry departments or, in the case of other operators, in financial incentives or restrictive covenants. A market is a necessary, but not a sufficient, condition.

Unfortunately all the arguments about raising the worth of forest products are potentially flawed. None of the measures proposed from time to time is a panacea; one has only to note how often reforestation funds have either been diverted to other purposes or used for purely cosmetic treatments. But, given will and control, such measures as the following would help greatly: higher prices, lesser-known species, non-timber forest products, differential royalties, more efficient capture of forest rent and its reinvestment in a reforestation fund, and financial confidence in forest management as an investment.

There should also be a direct relationship between the market and the degree of intensity of management that proves profitable. The management of natural forest has one potential advantage over intensive plantations: it is easier to play the market if the systems of management are not too rigid (this is what is done in effect in the long-standing forests of Europe). It will not be easy to do so, however, as long as the market is partially supplied by the first cut from virgin forest and from lands which are being cleared for agriculture at the same time as from forests under effective sustained-yield management – and in direct competition with them. In fact forestry suffers from operating within three distinctly different economic regimes: mining; sustained-yield management; and the "agricultural". It should exploit the strengths of natural-forest management: its flexibility, the quality of its products and the comparative advantage afforded by products that are well known and established in the trade. The alleged disadvantage, the long time of return, is mainly evident when the whole operation of a concession or a government forest is not turning over on a sustainable rotation, and if some of the initial profits derived from the first cut have not been reinvested in management.

Adequate information

We have emphasized the need for predictive systems providing the information both for policy-making and for management. To be cost-effective there must be careful choice of the correct kinds of information; immense amounts of energy and resources can be wasted in pursuing the wrong information or gathering data at the wrong level of precision.

Our analysis shows that there are critical deficiencies in knowledge in most countries concerning the resource, the likely market and much of the information required for management. Palmer (Chapter 6) provides a diagram of the critical knowledge required for sustainable management of a particular forest. On a wider canvas, national policy requires reliable information on: the extent of the forest; its present commercial volume; the growth potential of the forest; other claims on forest land (the traditional rights of forest peoples, the growing of food and cash crops, catchment protection, conservation, mineral exploitation, reservoirs, roads and human settlements). Adequate information for management and control requires, for example, inventories both for harvesting and for management, diagnostic sampling for regeneration, the results of measurements from permanent sample plots, and knowledge of the ecology of the principal economic species. There is also a need for accurate figures for the financial costs and returns from the whole range of activities associated with forest management, and with the harvesting and marketing of forest products.

To obtain all this information requires a timely and well-planned investment in survey, research and monitoring. Without it, it is not possible to exercise proper judgement in the selection of the permanent forest estate, or to provide a favourable financial environment for natural-forest management, or yet to establish the standards and criteria for controlling the forest operation. Once again this returns to the question of government resolve.

Conditions of success

To summarize, the conditions of success can be clearly identified as the following: government resolve to set aside a forest estate for the production of timber and to manage it sustainably; a sound political case for the selection of a permanent forest estate as part of a national land use policy; guaranteed security for the forest estate, once chosen; an assured and stable market for forest produce; adequate information for the selection of the forest estate and for planning and controlling its management; a flexible predictive system for planning and control based on reliable information about growth and yield; the resources and conditions needed for control; and the will needed by all concerned to accomplish effective control.

GENERAL CONCLUSIONS

This study has led us to some very firm general conclusions which have significant implications for the future of the tropical timber trade and for the management of natural forests in the producer countries of ITTO. These are likely to apply also to other countries with tropical forests which are not yet ITTO members. None of these conclusions is entirely new, but the nature of this study gives them, we hope, an authenticity and urgency that they have not possessed before.

We reiterate that the matter is of the greatest urgency. At present the extent of tropical moist forest which is being deliberately managed at an operational scale for the sustainable production of timber is, on a world scale, negligible.

Nevertheless, many countries have the intention, expressed in their legislation, to manage sustainably; and in a number of them partial forms of sustainable management are being practised either deliberately or by chance. Moreover there seems at last to be some awareness in most of the producer countries, and particularly among their foresters, that action is needed; and many countries, especially with the assistance of FAO and through the TFAP, are now taking some of the necessary steps to attain sustainable management. All the same, progress in establishing stable sustainable systems is still so slow as to have very little impact on the general decline in quantity and quality of the forest.

We are convinced that comprehensive and urgent measures are absolutely necessary if the tropical timber trade is to continue in the long term to handle material which even approaches the quantity and quality that it has become accustomed to. Even more important, the future existence of large areas of tropical forest, perhaps even the majority, and of the highly significant ancillary goods and services of the forest, depends equally on the establishment of sustainable systems of management, many of which must have timber production as their basis.

It has now been generally established, and is widely agreed, that the management of natural forest for the sustainable production of timber is technically possible in many forest types. Although more knowledge would be beneficial, the factors which constrain sustainable management are seldom principally technical.

In the course of this study we have identified a number of conditions, all of which must be present for sustainable management to succeed. These are: the establishment by government of a permanent forest estate within the natural forest with a guaranteed long-term future, as a part of an overall land-use plan; secure conditions for the managers of the forest, whoever they may be – government agency, private corporation, local community or other; the setting of standards for annual allowable cut, cutting cycles, tree marking, harvesting techniques, environmental safeguards, etc.; the adequate control of all aspects of harvesting and the treatment of the forest after harvesting to ensure that future crops are assured and that no unnecessary environmental damage is incurred; economic and financial policies which do not require more from the forest than it can yield sustainably (this requires a market, a government policy that treats forests as a resource to be managed not mined, and a reasonable distribution of revenues and profits among the various parties involved – government, managers, loggers, local communities, processors, exporters, etc.); and, finally, environmental policies which will satisfy a public which is becoming increasingly conscious of environmental problems.

Other factors necessary for the effective operation of all the above conditions include: clearly defined government objectives for the future of its forest lands and of timber as a commodity; information on the extent and quality of the forests, their soils, the quality of their timber, and their environmental values, to enable the firm definition of a permanent forest estate for timber production and the associated protection and conservation of forests; growth, yield and regeneration data from permanent sample plots which will enable models to be designed which can be used to determine the pattern of harvesting, the detailed marking of trees for felling and for retention, the silvicultural system to be applied, the length of cutting cycle and the nature of the future crop (the flexible use of such models will give the best chance of adapting in a sustainable manner the characteristics of the crop to the predicted market); accurate financial data covering all aspects of the operation; the best available forecasts of future markets and demand; and a knowledge of the environmental impact of timber operations and of the views of any local communities likely to be affected by them.

These conclusions and, particularly, the analysis of the necessary conditions for success lead to the recommendations for action which are the subject of the next chapter.

NOTES

1. These have been partly rewritten for two reasons: to be appropriate for the readers of a book rather than of an official report; and in order to clarify certain points in the original text which were misunderstood. These changes do not alter in any way the conclusions reached in the report or the general tenor of the recommendations.

2. FAO, in press, summarized in Schmidt, R.C., *Tropical Rain Forest Management: a Status Report* (Rome: FAO, 1987).
3. The scheme of extensive management being proposed in the Congo fits these criteria.
4. This is described in Thang, H.C., *Selective Management System: Concept and Practice (Peninsular Malaysia)* (Kuala Lumpur: Forestry Department, 1987). Although this is probably the best available system, it is still not even the complete theoretical answer, as it tends to make undue concessions to economic considerations and the convenience of concessionaires.
5. In 1980 the estimated total area of closed broadleaved tropical forest was 1,200 million ha of which 860 million were "productive". It was estimated that this area of productive forest would be reduced to 828 million by 1985. (Lanly, personal communication, 1988).
6. Wyatt-Smith, who was the author of the system, comments (personal communication) that the MUS was not solely a "silvicultural" system.
7. The area became a World Heritage Site under the World Heritage Convention in December 1988.
8. Burns, D., *Runway and Treadmill Deforestation*, IUCN/IIED Tropical Forestry Paper No. 2 (London: IUCN/IIED, 1986).
9. HIID, *The Case for Multiple Use Management of Tropical Hardwoods* (Cambridge, Mass.: Harvard Institute for International Development, 1988).
10. This may present difficulties where the state is on a joint-venture scheme with foreign interests as in some instances in Peninsular Malaysia.
11. Synott, Timothy, *Pre-project Report: Natural Forest Management for Sustainable Timber Production*, vol. 4 (London: IIED, 1988).
12. IIED/Government of Indonesia, *Forest Policies in Indonesia: the Sustainable Development of Forest Lands* (London: International Institute for Environment and Development, 1985).

8. Recommended Action
Duncan Poore

In this chapter we present the recommendations for action which
IIED made to the International Timber Council. The programme
recommended is, above all, designed to be practical and to promote
a rapid and comprehensive solution. All those who have been
occupied in this study are convinced that the time is past for tinkering
with the edges of the problem. There must now be wide-ranging and
comprehensive action.

It is important to realize, in considering these recommendations,
the limits within which they were drawn up: they are concerned with
the management of natural forest for the sustainable production of
timber, and they are directed to ITTO. The questions addressed are
only part of the much broader issue of the sustainable utilization
and conservation of tropical forest. Nevertheless, it is reasonable to
suppose that management for timber production should provide the
economic foundation on which the conservation of the tropical forest
must rest; and it is, therefore, significant enough to be considered
alone. We shall come back to some of the wider issues in the next
chapter.

THE ELEMENTS OF A STRATEGY FOR ITTO

The actions recommended below are derived directly from the
results of our study. They can, we suggest, be used to define the
elements of a strategy for ITTO in the sphere of natural-forest
management.

As a result of our study, we suggest that the elements necessary to establish on a firm, operational base the management of
natural forests for the sustainable production of timber are the
following:

- inspiring governments with a sense of urgency and purpose
- establishing in producer countries a permanent forest estate,

whose future is guaranteed, as part of a land-use plan which also provides adequately for protection and conservation forests

- developing the intellectual basis and justification for tropical-forest management for the sustainable production of timber at various different levels of intensity (as explained in Chapter 1); this would include consideration of the economic, social, institutional and environmental context of management at each level
- the wide promotion of such approaches in producer countries
- establishing standards and manuals of "best practice" for all the elements of management, harvesting and research in support of these
- encouraging proper control through adherence to these standards and practices
- promoting in the consumer countries an appreciation of the scale and complexity of the problem
- working through both producers and consumers to develop the best financial environment for sustainable production
- encouraging and financing the establishment of different models of management wherever the political environment is favourable; these models would be chosen so that their results might be of wide applicability or demonstrate innovative approaches
- encouraging an institutional environment favourable to consistent, long-term management for sustainable production by developing and encouraging appropriate training, policy guidelines, management structures, predictive modelling both of supply and demand, and research and monitoring systems
- encouraging co-operation between the environmental movement and the trade to create the conditions for sustainable management of properly chosen forests for production.

Not all of these actions need be carried out by ITTO itself; indeed the International Tropical Timber Agreement specifies that ITTO should, wherever possible, use the resources of other organizations in the field. Many of these actions, for example, are appropriate both to ITTO and to FAO, and it is clearly the responsibility of those governments which participate in the governance of both organizations to decide what the balance should be.

But there are certain kinds of action in which ITTO has a clear responsibility as a treaty organization of producers and consumers. These are issues of tropical-forest management which are concerned with the promotion, harmonization and sustaining of the trade flow in tropical timbers and the sustainable management of the forests upon which this trade depends. ITTO has a special role in issues in which producer and consumer nations have a common interest and those in which action will be especially promoted by interaction and contact between foresters, forest industries, exporters, importers and final users. By nature of its special provisions it also provides a profitable forum for exchange of views on environmental aspects of natural-forest management.

In this connection, we would argue that all the issues that we have raised are so important for the future of the tropical timber trade that ITTO, even if it does not implement them itself, should play an active part in inducing others to do so!

RECOMMENDED ACTIONS

When we considered the actions needed, we found that they could be divided logically into four groups according to the manner in which they would contribute towards the widest possible extension of effective sustainable forest management.

First, there are actions of *promotion* – those that are aimed at all those engaged in natural-forest management or affected by it. Next are actions of *diagnosis* – concerned with the further examination of critical issues to determine exactly where action is most needed and what that action should be. The third group is concerned with *providing examples*. This includes two elements: to give the greatest publicity to existing models of successful operation so that they may be applied elsewhere and used in training; and, where necessary, to develop new models. The last groups can best be called *facilitation*; the actions included in it are devoted to providing any necessary aids to make the large-scale expansion of successful management easier, more rapid and more effective.

The following, then, are the actions recommended grouped in these four categories.

Promotion

Whether natural-forest management for sustainable timber production becomes a reality depends fundamentally on the producer countries themselves, for all the elements that we have identified are primarily their responsibility: the security of the forest estate depends upon land-use policy, legislation and control; satisfactory management depends upon effective monitoring and control, based upon the results of domestic research; many, though not all, of the financial conditions depend on national economic and fiscal policies and their enforcement; the information required depends upon the definition of research priorities; and all depend upon the availability of enough qualified, honest staff who receive adequate remuneration for their work.

Past experience suggests that action on this scale will require a great change in attitudes towards the utilization of the natural forest on the part of governments and influential public opinion in the producer countries. The message of this book is, therefore, addressed to them before all others. Without their real commitment to making enough resources available for the truly sustainable management of their forest lands, nothing that can be done by outside agents will save the situation. Fortunately, there is some indication in many countries that people are becoming aware of the nature and extent of the problem. The role of the international community is primarily to make this task easier for them.

The challenge, therefore, is to create in the producer nations the milieu in which sustainable management can succeed. Both financial and technical assistance are, of course, important in accomplishing this. But perhaps of greater importance is the creation of a sympathetic international environment. We suggest that ultimately the two most influential agents in this process will be the financial and the environmental. The timber trade and importing governments have an important role to play in the former; and, in the latter, those non-governmental organizations who have been vocal critics of recent tropical-forest management.

We have found that there are many misconceptions which hinder the management of natural forest for sustainable timber production. Among the subjects which are misunderstood are: the extent and urgency of the problem; the nature of the actual constraints on operational forest management; the fact that sustainable management is technically feasible; the real values of forest lands and of the timber and other goods and services from the forest; the limited benefits of converting forest to other uses; and the relative benefits and costs of obtaining timber from natural forest rather than plantations.

These messages need to reach many different audiences. The most important of all are the controlling centres of government, especially perhaps heads of state and their advisers, and ministries of finance and of planning. But it should also reach the trade and all concerned with it, competing interests for forest land, financial institutions, public opinion in the producer countries, external organizations who provide finance and technical assistance, and environmental groups. A large and comprehensive exercise in public relations and persuasion is required to reach these groups.

Many of the communities who live in or near the forest also need to be persuaded of the value of retaining the forest as forest rather than converting it to other uses. They are more likely to be persuaded by demonstration and example.

There are many actions which could be recommended here. We have picked out a few that we consider could be influential.[1] Two are concerned with marshalling general arguments that must be used to convince senior politicians and the planning ministries of governments: developing the economic case for natural-forest management (Action 3), and presenting the arguments for the quantity of tropical forest needed by the countries in which it occurs (Action 2); both of these are vital if a convincing case is to be made for forestry *vis-à-vis* agriculture. One is designed to bring to the attention of governments the ways in which the interaction of their policies affects the future of the forest (Action 1). The last (Action 4) suggests a number of meetings in which all these issues could be given the maximum exposure among influential groups of people. Short explanations of these follow.

1. Policy reviews embracing the forest sector
There should be policy reviews in each of the producer countries in which special attention is paid to the matters raised in this book, for example: the interaction of national economic and land-use policies on the nation's forests; the economic worth of the forest, and the direct and indirect contribution of forestry to the nation's wealth; the place of sustainable natural-forest management in providing for the needs of domestic and export markets; the need to define and protect a permanent forest estate; the claims of this forest estate in competition with other forms of land use and the effect of other government policies on the prospects for retention of land under forest use; the need for strong government backing to enable all aspects of harvesting and forest management to come under adequate control; fair distribution of rents and profits from forest enterprises; and innovative ways of ensuring that those who are given responsibility for managing the

forest have a direct interest in managing it well. The TFAP might well be the agent for such reviews provided that its approach is wide enough to embrace these policy interactions; indeed such broad reviews have taken place in a number of countries.[2]

2. Study of the amount of tropical forest that must be retained

An authoritative study should be carried out of the amount of tropical moist forest required by the countries in which it occurs. What is envisaged is a study similar to FAO's *Land, Food and People*[3] to provide firm and convincing evidence upon which the case for the retention of a substantial area of natural forest might be argued. This would look at the area of land (and its location) required for the production of timber, fuelwood and other forest products, catchment protection, the protection of other critical forest soils and habitats and the conservation of biological diversity.

3. The economic case for natural-forest management

The full economic arguments for natural-forest management should be assembled in order to publicize and promote them to those audiences with the greatest potential influence: to economic summits, regional policy meetings such as those of the OECD, OAU and the Commonwealth prime ministers, the councils of the international development banks and other financial institutions, national economic and planning ministries and the media.

The study should build upon the material developed, for example, in the HIID pre-project and in the studies collated in Repetto and Gillis.[4] It should include comprehensive assessments of the value of the services provided by the forests: option values; detailed assessments of the economic viability of the management of natural forest for sustainable timber production at various levels of intensity and in relation to future projections of possible timber demand and prices; the issue of comparative advantage over plantation species; and the supplementary contribution that might be made from non-timber goods.

The output should be reports directed to governments, books, material for the media, etc. It should provide an important input to policy discussions such as those advocated in Action 4.

4. ITTO-sponsored meetings

There are a number of possibilities here, all designed to heighten the importance of natural-forest management in the perception of those who determine policies. The following are suggested as possibilities.

(a) Regional meetings of producer countries designed to secure the financial and institutional support needed for sustainable forest management: they would therefore involve the highest levels of the ministries of finance and planning.

These meetings should look at all aspects of policy affecting the natural forest, with particular reference to production forests (but also including protection and conservation forests), and would cover: the needs of countries in the region for tropical hardwood timber both for domestic use and for export; the place of timber (and other forest goods and services) in domestic and regional economic planning; the consequent need for a secure and permanent forest estate and the implications of this for other potentially competing uses of the land; the possible size and location of this permanent forest estate; and identification of the needs for investment in forest management and in the institutions to support it.

The meetings should include representatives from the highest levels of the ministries of finance and planning, from forest authorities, and exporters and importers of tropical timber; forestry and timber trade associations; forest industries engaged in extraction, conversion and manufacturing; financial institutions; representatives of forest communities and of environmental interests.

(b) Meetings at a planning and technical level of blocks of producer and consumer countries which are substantial trading partners in tropical timber: these would be designed to examine the supply and demand prospects for their bilateral trade against the background of the present state of the resource, and to study alternatives for the future based on combinations of natural-forest management and supplies from plantations, agroforestry, etc. These should include representatives of the forest authorities, exporters, importers and forest industries.

(c) A special meeting, perhaps jointly organized by ITTO and FAO, looking at the extent to which the TFAP has succeeded in addressing the question of sustainable management of tropical moist forest in the course of its sector reviews, what new investment is being provided as a result, and what further action should be taken.

(d) A high level meeting addressed to the issue of the future of natural forest in the tropics: this might be in association with, or preceded by, the meeting suggested above, (c).

Diagnosis

It is a frequent refuge for those who do not wish to take action to recommend more research or investigation. Nevertheless, there are a number of areas in which more information is urgently required if the right actions are to be taken to promote the sustainable management of natural forests. Some examples are given below.

5. Tropical forest resource assessments

This study has found that there is still very great uncertainty about the areas and status of the forests in many of the countries visited. This situation should be remedied by the tropical forest resources assessment to be carried out by FAO based on the position in 1990.[5] It is vital that reliable information should be available for ITTO as soon as possible.

Natural Forests In terms of ITTO and, indeed, FAO interests, it would seem that special care needs to be taken in the very difficult task of assessing the management status, both legal and actual, of natural-forest areas. Very great changes can occur in the status of non-virgin, modified forest according to the intensity of interference or the effectiveness of control. Logged forest which is allegedly under management may in fact be degrading so rapidly that it is effectively unproductive, while abandoned cultivation or grazing lands may rapidly become colonized by potentially valuable trees. (The potential importance of degraded forest lands in this respect is now being recognized, especially in Latin America.)

For ITTO it is most important to have detailed information, country by country, of those categories which have a direct relevance to future timber production. Some of the most significant facts required (and some of the most difficult to obtain) are those concerning the security (both legal and actual) of the potentially productive natural-forest estate and the real degree of management control. Without this information any figures can prove very misleading.

Plantations Associated with the plantation segment of the FAO tropical forest resources assessment should be a desk study of the possible contribution to the trade in tropical timber likely to be made by hardwood plantations. This should look at: the actual and potential area of such plantations; the nature and quality of the product from them; the likely influence on the existing tropical timber trade of the access to it of plantation-grown timber; and the probable interactions between investment in plantations and investment in the management of natural forests.

The relative role of plantations and natural forest is likely to be one of the most significant issues for debate in the future forest policies of the producer countries – whether to invest in plantations ("timber estates") or in forest management. Much more exact and detailed knowledge is required about the interactions between the two before it will be possible to make rational decisions about a reasonable allocation of resources between them.

6. Tropical timbers: financial aspects of harvesting, management and trade
An assessment should be made of the financial structure of all the stages involved in the growth, harvesting, transport, transformation, export and import of tropical timber. The study of royalties would be an important part of this and should look at the situation in each of the producer countries and their marketing partners in the consumer countries. It should attempt to assess the costs of the various stages in dealing with timber and the distribution of rents and profits to the various parts of the enterprise and the various partners in it.

By bringing a degree of transparency and comparative analysis to this financial structure, it should provide evidence to assess a whole number of questions that are now obscure, for example: the cost-effectiveness of natural-forest management as compared with the establishment of new plantations; the proportion of "rent" which is recovered by producer countries and the amount that is reinvested in forest management; the likely effect of increasing the prices paid for tropical timbers; the level of royalties and the effect of differential royalties; the possible effects of bringing on to the market a number of "lesser-known species"; the real cost to producer countries of managing their forests in a sustainable manner; whether unreasonable profits are made at any stages in the process, etc. These obscure areas seem to the authors to be among the main constraints inhibiting the wider introduction of sustainable management to natural forests.

Such an assessment might be carried out as an ITTO pre-project arising directly out of those carried out by HIID and IIED. It could lead to the development of computer modelling of the systems involved and the possible effects of modifying them.

The ultimate objective would be to design and promote the adoption of financial structures that provided reasonable profits for all partners and allowed sufficient sums for reinvestment in forest management.

7. Identification of field projects where sustainable production of timber and non-timber products may be combined

The purpose would be to identify actual situations where, as in Acre, Brazil (with brazil nuts and natural rubber), projects might be developed which are based on the economic harvesting of both timber and non-timber products.

The theoretical case has been made in the HIID study for supplementing the production of timber with that of non-timber products because the addition of these non-timber products may increase the economic benefits to be derived from harvesting, and may make the operation more socially acceptable. Both of these lines need to be explored further. IUCN is involved in a number of projects which are designed to increase social acceptability; it should be encouraged to do more. But there are still few examples of actual sites where it might be possible to increase the market benefits from an area of forest by including non-timber products. This is largely, it seems, because the non-timber products rarely enter the organized market. Any promising example should be identified and promoted. This could be handled by an ITTO pre-project.

8. Examination of possible ways in which consumers might encourage producers to adopt sustainable management

A number of measures have been suggested whereby consumers might influence producers to adopt high standards of sustainable management. Among these are: quality marking of products which come from forests which are managed for sustainable production; awards for good management; and black-listing for bad management. These schemes would not be easy to operate and it is by no means certain that they would all have the desired effect. It is suggested that the various proposals should be examined critically to determine whether they are desirable or practicable. If they prove to be so, they should be actively promoted.

Examples

The general opinion, expressed in our many discussions in the countries and in the regional seminars, is that there is a need in many countries for more projects, funded by governments themselves or assisted by any appropriate agency, which demonstrate the feasibility of natural-forest management for the sustainable production of timber. There is a preference for projects which occur in situations, both ecological and social, that are repeated elsewhere so that their results can be used for demonstration and training, and can be replicated, and which illustrate possible new models for sustainable management, such as those operated by agencies of government at different levels, by private corporations or by local communities. Strong political backing is also an essential condition.

It is recommended that most of the field project activities of ITTO concerned with reforestation and forest management should have these characteristics. Within these conditions, the more demonstration projects that are developed the better. Projects should always include provision for training and publicity and for receiving visitors from other members of ITTO.

9. *Innovative models of management*

Projects should be encouraged which develop models of management for the sustainable production of timber that are innovative but capable of being widely adopted either within the region or in comparable circumstances elsewhere.

Possible examples are those: (a) that involve the management of forests by private enterprises or forest industries, and which involve the use of private finance; (b) where the whole management of the forest is devolved to levels of government that are lower than the national; (c) where, as in Honduras or Gabon, the local community is the agent of management.

Facilitation

Certain services could be provided by ITTO (or other agencies) which would make it easier for producer countries to develop rapidly the capacity to apply sustainable management widely at an operational level. The following are the most important suggestions that have arisen from this study and from the regional seminars.

10. Guidelines of "best practice"

Enough experience has now been gained in various parts of the tropics for it to be possible to draw up manuals setting out "best practice" in relation to most of the activities involved in the sustainable management of natural forest for timber production.

It should be the aim of ITTO to ensure that a comprehensive series of manuals is produced covering such subjects as: the definition and establishment of a permanent forest estate; the choice of a silvicultural system; cutting cycle analysis; tree marking; methods of harvesting and their control, especially to reduce damage to valuable residuals; the design and construction of roads; environmental considerations in management for timber production; and financial considerations in the management of natural forest.

Once these have been produced, they should be discussed within the industry in order that they may be adopted in practice as widely as possible. Once this stage is reached, various incentives might be offered to those operators who follow the practices recommended.

11. Guidelines for designing a research programme to give sufficient information for yield control

Much work has now been done, notably in Malaysia, Queensland and Surinam, on the data which are required to produce sufficiently reliable models of stand dynamics to determine allowable cuts, to mark trees for felling and retention, and to determine the desirable length of cutting cycle. If such models are used in a flexible and sensitive manner and there is strict control of operations, it should be possible to continue to manage sustainably (see Chapter 6).

A manual should be produced on the steps that need to be taken in order to manage the forest sustainably with minimum intervention, and to assess the effects on yield of post-harvesting treatments including enrichment planting.

In the first instance this might be confined to the variables needed for silvicultural control. But, because sustainable management depends also upon financial variables, the models might be expanded to cover the income and expenditure involved in all aspects of forest management and timber extraction.

12. Permanent sample plots

As part of the assistance required to establish systems of sustainable management, encouragement and assistance should be given to producer countries in establishing, maintaining and using an adequate network of permanent sample plots for determination of yield and regeneration potential. These should be established according to approved patterns and ITTO should maintain a central register of them. At the same time an attempt should be made to identify and bring back into operation any plots which have been neglected but which could still produce useful results.

13. *Appropriate training*
There should be co-operation between ITTO and the Forestry Education and Training Branch of FAO to draw up specifications for the kinds of training necessary to ensure: that forestry departments in producer countries contain trained staff who are able to argue with full confidence and authority the economic, social and environmental case for a permanent forest estate and who are competent in all necessary aspects of natural-forest management for sustainable production of timber and other products.

This does not mean that all staff need to be trained in everything, but it does mean that there should be staff suitable for carrying out the very varied tasks required of a forest department. For example, there must be headquarters staff with the knowledge and ability, in economics, general land use and social matters, to present to ministries of economic affairs a convincing case for resources for forestry. At field level there must be staff who have enough knowledge of rainforest ecology to be able to respond flexibly and intelligently to situations that they find in the field, and who have understanding of the problems of logging and forest industries and a sympathetic approach towards local peoples.

14. *Tropical-forest management information network*
There should be an investigation of the feasibility of establishing an information network at the working level among those in the three continents who are concerned with natural-forest management to ensure that the latest thinking and experience get the widest possible currency.

We were impressed, in each of the continents, by the great need to establish a means of disseminating information, especially up-to-date and unpublished information, about tropical-forest management among those who are concerned with the subject in the field. This point was made in all three regional seminars. There is a special need for dissemination of material between those of different languages; for we have found that there is very little interchange of information between those using English, French and Spanish, far less between those who use other working languages.

15. Exchange visits

A programme of exchange visits for technical personnel is desirable in order that they may benefit from the most recent experience in other countries. This should be linked with the demonstration projects which have been suggested above.

NOTES

1. It has been suggested that one of the important actions should be to prevent any removal of forest for the purpose of encouraging (or conniving at) the extension of agriculture, for there is no doubt that more forest is destroyed in this way than by any other means. This is without any doubt a very critical issue which may perhaps best be addressed indirectly. Actions 1-4 were all formulated with this in mind.

2. The reviews carried out by IIED in Indonesia, Cameroon and Zaïre provide examples of the approach. A very important component would be the kind of information collected in Robert Repetto and Malcolm Gillis (eds), *Public Policy and Misuse of Forest Resources* (Cambridge: Cambridge University Press, 1988).

3. FAO, *Land Food and People* (Rome, 1984): based on the FAO/UNFPA/IIASA report *Potential Population – Supporting Capacities of Lands in the Developing World*.

4. HIID, *The Case for Multiple Use Management of Tropical Hardwood Forests* (Cambridge, Mass.: Harvard Institute for International Development, 1988); Repetto and Gillis, op. cit.

5. This is referred to in two papers published by FAO in 1987: *Report of the ECE/FAO/FINNIDA* ad hoc *Meeting on Forest Resources Assessment*; and *Concepts and Methodology of the Tropical Forest Resources Assessment (1990)*, Paper prepared by K.D. Singh for the ECE/FAO/FINNIDA Forest Resource Assessment Meeting, Kotka, Finland, 26-30 October 1987.

9. Recent Developments
Duncan Poore

The year 1988–9 has been one of great activity on the tropical forest front. An account is given in this chapter of some of the more important recent developments.

ITTO: SUBSTANTIAL PROGRESS

In the last year ITTO has made substantial progress in developing its very special role, although its activities are still severely curtailed by the failure of many members to make their statutory contributions to the administrative account. The situation may shortly be eased by a large contribution from the government of Japan towards the finances of the organization. Panama has now acceded to the agreement, the nineteenth producer country to become a member.

The report prepared by IIED was presented to the Fifth Session of the International Tropical Timber Council at its meeting in Yokohama in November 1988. At the same meeting the council established a panel in order "to review the current development situation of tropical forests and prepare a draft action plan for ITTO in the field of reforestation and forest management and present this to the Permanent Committee on Reforestation and Forest Management". In doing this they were charged to take into account this study, the study by HIID,[1] and two others concerned with the establishment of plantations and with enrichment planting prepared by the Japanese Overseas Forestry Consultant Association (JOFCA). The report of the panel was presented to the Sixth Session of the council which met from 16 to 24 May 1989 in Abidjan. It was based strongly on the detailed recommendations of the IIED and HIID reports, and contained a list of twenty-one proposed actions.

The subject was vigorously debated at Abidjan, being considered by all three committees of ITTO, and it formed the subject of a Resolution of the Council, the operative parts of which read as follows:

Requests the Executive Director to prepare a work programme of the Secretariat for the years 1990 and 1991 in the field of reforestation and forest management based on the proposals made in [the report of the Panel] and on comments made on these proposals.

Further requests the Executive Director to propose activities for priority implementation in the following areas of action:

- Develop guidelines for "best practice" and sustainability in relation to the sustainable management of natural forests and plantations;
- Develop the economic case for natural-forest management;
- Strengthen policy initiatives embracing the forestry sector;
- Increase awareness and mobilize support to ensure the sustainable management and conservation of tropical forests;
- Develop demonstration models of management for the sustainable production of timber and non-timber products and conservation;
- Strengthen research on responses to silvicultural treatment;
- Develop human resources in tropical forest management;
- Consider incentives to encourage sustainable forest management.

Requests the Secretariat to prepare elements for an Action Plan in the areas of forest industry and economic information and market intelligence supported by a panel of experts in forest industry to be submitted to the next Sessions of the Permanent Committees and of the Council.

Decides to include this item on the agenda of the next Session of each of the Permanent Committees and of the Council with a view to adopting a comprehensive action plan to serve as a basis for the work programme of the Organization and the Secretariat.

This meeting of the council gave considerable encouragement to those who believe that sustainable management of natural forest for the production of timber is one of the keys to forest conservation and to the timber trade. Although it was a disappointment to some that the Action Plan was not approved as a whole, the firmness of the Resolution gives good grounds for optimism. Added to this, member governments, perhaps encouraged by this progress, pledged approximately another $8.5 million to the Special Account and agreed to extend the Agreement for a further two years. Two important demonstration projects should also begin shortly, both in areas mentioned earlier in this book: the integrated management of forest in the state of Acre, Brazil, and a comprehensive project in Chimanes, Bolivia. Work has started, too, on planning the rehabilitation of the burnt forest area in Kalimantan, Borneo.

The most exciting and potentially influential decision of the council was, however, unanticipated – and unprecedented. A generous, courageous and imaginative invitation was issued by the government of Malaysia and the Chief Minister of Sarawak to enable a mission to visit Sarawak to examine the status of natural-forest management. This invitation was accepted by the council. Because of its importance, the Resolution adopted is quoted in full.

Resolution 1(VI)
The Promotion of Sustainable Forest Management: a case study in Sarawak, Malaysia

The International Tropical Timber Council,
Reaffirming the obligation and commitment of all Members to the objectives of the ITTO, 1983,
Bearing in mind Article 1(a) of the ITTA, "to provide an effective framework for co-operation and consultation between tropical timber producing and consuming Members with regard to all relevant aspects of the tropical timber economy", and Article 1(h) of the ITTA, "to encourage the development of national policies aimed at sustainable utilization and conservation of tropical forests and their genetic resources, and at maintaining the ecological balance in the regions concerned",
Recalling the Statement made by the Representative of Malaysia at the Fifth Session of the International Tropical Timber Council informing the Council of the serious efforts to promote sustainable forest management in Malaysia and inviting international assistance to support the implementation of these policies,
Taking note of the Statement made by the representative of Malaysia at its current Session of the ITTC,
Expressing its appreciation to the Government of Malaysia for its readiness to welcome a Mission to visit Sarawak, Malaysia at a date to be decided by mutual agreement,
1. *Establishes* a Mission with the following terms of reference:
    a) To assess the sustainable utilization and conservation of tropical forests and their genetic resources as well as the

maintenance of the ecological balance in Sarawak, Malaysia, taking fully into account the need for proper and effective conservation and development of tropical timber forests with a view to ensuring their optimum utilization while maintaining the ecological balance, in the light of recent ITTO studies on forest management for sustainable timber production in Member countries and relevant reports by other organizations;

b) Based on its findings, to make recommendations for further strengthening of sustainable forest management policies and practices, including areas of international co-operation and assistance.

2. *Authorizes* financing not exceeding $300,000 from the Pre-Project Sub-Account for the work of the Mission.

3. *Appreciates* the readiness of the Government of Malaysia to fully co-operate in facilitating the work of the Mission and to allow it to visit any part of Sarawak, to meet any persons and also to make available information relevant to the work of the Mission.

4. *Invites* all Members and relevant international organizations and international institutions to lend their fullest support for the success of the Mission.

5. *Appeals* to all Members, bearing in mind Article 30 of the ITTA, to use their best endeavours to co-operate to promote the attainment of the objectives of the ITTA and avoid any action contrary thereto.

6. *Requests* the Executive Director to take all necessary measures for the implementation of this Resolution and to prepare the necessary documentation for this purpose.

7. *Requests* the Executive Director to communicate this Resolution to all international organizations and others interested in the work of the ITTO.

8. *Further requests* the Mission to present, on a confidential basis, a Progress Report at the Seventh Session and its final Report at the Eighth Session.

This remarkable international initiative demonstrates the special ways in which ITTO can draw together consumer and producer nations in the interests of sustainable management in a way which has hitherto proved impossible outside the framework of this agreement.

The success of the agreement will be judged by the extent to which: producing countries, through its influence and help, succeed in establishing policies for the sustainable management and conservation of their forests and for the sustainable production of wood; the trade in tropical timber becomes more stable with an increased flow of benefits, equitably distributed between the parties concerned.

TFAP: CHANGE OF GEAR

The other great international effort to stem tropical deforestation and promote the conservation and sustainable development of forest lands in the tropics is the Tropical Forest Action Plan which came into being in 1985. The main thrust of this has been to assist governments in reassessing their priorities in forest development, to increase the amount of international investment in forestry, and to co-ordinate the responses of the main multilateral and bilateral donors to requests for aid and technical assistance. The plan received endorsement from the World Forestry Congress meeting in Mexico City in June 1985. Since then it has operated mainly by carrying out review missions of the forest sector in the many countries that have requested them, missions which are designed to help countries examine their priorities in the forestry sector and to draw up projects for new investment. The work is co-ordinated by a Co-ordinating Unit within the Forestry Division of FAO in Rome and is guided by biennial meetings of the TFAP Forestry Advisers Group whose working document states that it is

> an informal assemblage of senior forestry advisers dedicated
> to more effective coordination, cooperation and collaboration
> between donor agencies, developing countries and organizations
> concerned to improve the availability and use of human and
> financial resources for the conservation and sustainable develop-
> ment of tropical forest resources.

The first missions promoted by the TFAP started in 1986 although some similar missions that had begun earlier were absorbed into the TFAP process. Action plans have now been adopted in ten tropical countries and forty more are being prepared. Among ITTO members, review missions have been carried out in Cameroon, Côte d'Ivoire, Ghana, Honduras (conducted nationally), Panama and Peru; and reviews are taking place in the Congo, Ecuador, Indonesia (national), Malaysia (national), Papua-New Guinea, the Philippines and the CARICOM islands, including Trinidad and Tobago. The whole process is necessarily somewhat slow and new investments have only just begun. It is rather early therefore to assess whether this approach has had a dramatic effect in changing priorities. There is no doubt, however, that it has approximately doubled the amount of money devoted to forestry by aid agencies.

The TFAP was originally criticized on a number of counts. First it was thought that a process which reviewed the forestry sector alone was too narrow, when most of the problems of deforestation and wood shortage had much more complex origins in the policies of many different parts of government, especially in agriculture and in economic and physical planning. A second criticism was that it laid too much stress on the power of investment from outside rather than on changes of policies and priorities within governments themselves. Other points concerned the operation of the Advisers Group as a "club" of donors and the lack of attention paid to NGOs and community groups in the course of missions. There has clearly been validity in some of these criticisms; others have been based on misunderstandings of how the TFAP worked. A number of them have been reiterated recently as a result of a consultation carried out by the World Resources Institute in 1989.

The Forestry Advisers Group and the Co-ordinating Unit recognize these points and are responding to them. As experience grows, the approach of the missions becomes more comprehensive and more sensitive. The intentions come out clearly from the "basic principles which characterize a TFAP strategy" which were discussed at an Advisers' Meeting in Paris in May, 1989.

- A declared political commitment at higher government level to forestry and its role in rural development.
- Forestry policies which reflect the process of reconciliation of present and future needs for sustainable development and the environmental role of forests and trees.
- A declared focus of forestry policies towards meeting the needs of local people particularly the rural poor who depend on forest and tree resources for their subsistence and food security, and as a major opportunity for development.
- A visible role for forestry in national development plans with a clear indication of objectives, priorities and allocation of increased public resources.
- Active, organized and self-governed involvement of local groups

and communities in forestry activities, with a particular focus on the most vulnerable and on women and on commonly shared resources.

- The involvement of rural people, local NGOs and the private sector in the planning and management of forestry activities.
- An identification of the main constraints and problem areas requiring immediate action to restore or maintain the critical role of forestry in environmental stability and socioeconomic development.
- A systematic combination of actions geared at monitoring and conserving the resource base and at raising and broadening the goods and services produced by forests and trees.
- Effective co-ordination of policy, planning and implementation of activities among relevant departments involved in the primary use of land such as agriculture, livestock, forestry, mining, energy, etc., and also with those involved in processing (cottage and industry) and commerce.
- Increased public and private, national and international investments to increase the production of goods and services from forestry.
- An effective and increased support by the international community based on a concerted response to technical and financial assistance needs and priorities expressed by tropical countries in line with the principles of TFAP.

ACTIONS BY PRODUCER GOVERNMENTS

A number of steps have been taken in the last twelve months by producer countries which demonstrate clearly that they recognize the seriousness of the situation and the need to take action, although in some cases the exact effect is far from clear. In January 1989, Thailand, for example, banned any further logging by cancelling 301 logging concessions; it had been provoked into this action by serious floods in November 1988. The Philippines has banned the export not only of round wood but of sawn timber also. Important revisions of the conditions of concessions are taking place in both Indonesia and the Philippines, and Indonesia is reviewing the effectiveness of the Indonesian Selection System (TPI). In Brazil, the IBDF has recently been disbanded and most of its functions have been transferred to the new Brazilian Institute for the Environment and Renewable Natural Resources. Cameroon is embarking on a programme of detailed resource planning for the extensive forests in the south east of the country.

All of these may portend more effective measures to bring forests under sustained management. Only time will tell. In some countries they are certainly a rearguard action in response to a situation which has already become critical; in Thailand, for example, the forest area has been reduced from about 65 per cent of the total land area twenty years ago to about 12 per cent at present. In others such as Cameroon and Brazil, it may represent a timely effort to plan for the use of natural forest before it is too late.

ACTION BY THE TIMBER TRADE

The timber trade was rather slow to react to the possible threat to its sources of supply. It was, for example, poorly represented at early meetings of ITTO. This situation is now changing. A spokesman of the UK hardwood trade is reported as saying recently:[2]

> The message is clear. You may think you have lots of forests, but one day you find you haven't. And at that stage, not only do the rewards of the timber industry suddenly start to dwindle, but so the realization grows that it's going to take a long time and a lot of money and effort to put the resource back in place. It may even be too late.

An important initiative was taken in the summer of 1988 by the Nederland Houtbond and the UK Timber Trade Federation which led, at the seminar held in Yokohama in November, to a statement by T.S. Mallinson on behalf of the Union pour le Commerce des Bois Tropicaux (UCBT) of the EEC.

He identified at least seven factors which the trade could see in the search for sustainable management:

1. To secure the future the forest must be controlled and managed.
2. Logging operations must be controlled.
3. Governments should be persuaded to control land clearance and migration into the forest which result in its destruction.
4. Research and educational programmes are essential to prove to local populations that their forest is their capital asset and that the products from that forest provide the interest on their investment.
5. Practical research in the areas of forestry, environment, logging and industrial development will be needed.

6. The existence of markets for forest products must encourage responsibility for the resource.
7. Taxes and royalties raised in the producer countries from the forest products sector should be applied within the sector.

He went on to discuss the few possible options to alleviate the problems in tropical forested countries. These would be: the total or partial banning of timber exports by the producer countries; the total or partial banning of imports of tropical timber by the consuming countries; applying limitations to the use of wood within the producer countries; improved reforestation and forest management financed entirely by producer countries; further strengthening of sustainable forest management and related activities by direct contribution from the trade and industries of the consumer countries. There is another option: to let the present situation continue.

This analysis led to the introduction of the concept, considered by the board of the UCBT,

> of a surcharge to be raised on imports of tropical timber products by the European Community and, generally speaking, by the world, from whatever source. . . . It would be logical to see it put into a special fund and channelled through the mechanism of ITTO into activities strongly focused on the structures and mechanisms of forest management.

This whole question of a surcharge is now being widely discussed. There are concerns among the tropical nations that the cost might be passed on ultimately to the producer, however good the intentions that it should be otherwise; and that this might render tropical timber less competitive in the international market and make it more liable to substitution. Among the consumer nations there are technical questions about the relations of the proposed surcharge to other tariff and non-tariff barriers – its conformity, for example, to EEC rules and GATT – about the ways in which the surcharge might be collected and about the best ways in which it might be disbursed. These discussions are likely to be protracted. If all technical problems can be resolved, however, such a direct connection between the trade in the consumer countries and the better management of tropical forests is highly to be desired.

CONSUMER CAMPAIGNS

During the past few years a great deal of effort has been put by environmental groups, especially the Friends of the Earth, into raising awareness among timber users that there is an ecological cost to the use of tropical hardwoods. There is little doubt that this has been successful in raising awareness and has played some part in stimulating the timber trade to make its proposals for the surcharge. In their extreme form these campaigns have called for the banning of the import of all tropical hardwoods. But it has generally now been accepted that any complete ban could be very harmful. So attention has shifted towards a policy of discouraging the use of timber from forests that are not managed sustainably and proposing a system of labelling of products from well-managed forest.

A significant number of consumers are reacting to these proposals. In the UK several national retail chains and a number of smaller independent companies have agreed to withdraw products containing tropical timber unless it can be shown to have come from a sustainable source; and more than thirty local authorities have adopted a policy of ceasing to use tropical hardwoods. Associations of traders in the timber and building industries have been established with the purpose of promoting awareness and taking appropriate action on issues of tropical timber and rain-forest destruction. In Holland 40 per cent of local governments have decided to reduce the use of tropical hardwoods. Similar actions are being taken in Belgium and the Federal Republic of Germany. On 26 May 1989 a Resolution was passed by the European Parliament referring to the regulation of European Community trade in tropical timber, the main elements being:

- The establishment of a system of import quotas by bilateral agreements between the EC and tropical timber-producing countries.
- To base the quota system on sustainable production levels of the producing countries.
- To increase financial and technical assistance by the EC in order to help in implementing the quota system and to compensate tropical countries for any decreases in foreign exchange earnings.[2]

Similar thinking is being developed in the United States, as is illustrated by the following resolution adopted at a workshop held in the middle of April 1989 in New York:

Resolution of the Workshop on the United States Tropical Timber Trade

We the undersigned, having gathered in New York City on April 14-15, 1989 for the aforementioned Workshop, and representing US industry, conservation organizations, academia and others concerned with forest conservation and the sustainable supply of tropical timber, do hereby resolve:

That a boycott of tropical timber products, except in specific circumstances, would not promote sustainable forestry in producing countries at this time;

That the International Tropical Timber Organization (ITTO) and the Food and Agriculture Organization of the United Nations (FAO), in consultation with other relevant agencies (including but not limited to bilateral and multilateral development agencies), should devise and promote an internationally-agreed-upon system for rating and documenting tropical timber production according to sustainability, and publicize these results;

That all concerned parties should recognize the ITTO as a key international forum for promoting improved forest management practices by industry and regional coordination of forest policies, and help to fund and otherwise participate in the ITTO;

That all countries should provide meaningful funding to support the work of the ITTO; in addition, the US participants present specifically request that their government fund the ITTO at a significantly increased level – at least five to ten million dollars annually;

That all relevant agencies should encourage pricing policies that assign to standing timber its correct stumpage value;

That all relevant agencies should devise strategies to determine, assign and capture the real value of non-timber goods and services (environmental services, biodiversity, "minor" forest products, etc.), beginning immediately and incorporating research findings as they become available;

That the rights of native forest dwellers should be recognized in land use planning processes to determine areas of permanent forest estate, protected forest, production forest and non-forest uses;

And that all concerned parties increase their support for model projects on sustainable forestry.

Lastly, we resolve to promote these aims, for the purpose of which we have formed a Committee duly representative of those present.

NOTES

1. HIID, *The Case for Multiple Use Management of Tropical Hardwood Forests* (Cambridge, Mass.: Harvard Institute for International Development, 1988).
2. *Tropical Rainforest Times*, Spring 1989 (Friends of the Earth, London). This journal, and *Eco*, the NGO newspaper produced for the International Tropical Timber Organization, are the source of this quotation and of topical news of developments in the field.

10. Postscript: an Uncertain Future
Duncan Poore

There is no simple basis on which to decide the balance between preventive and therapeutic measures. It is a problem that confronts society in many fields, and reformers, if they are to be trusted, need to feel for the present and think for the future. We cannot respect the judgement of people who are so preoccupied with the present that they give no thought to what is to follow; but we cannot trust the motives of those who can ignore existing problems when designing a future blueprint.

(Thomas McKeown, The Origins of Human Disease, Blackwells 1988).

The many different currents in the present flow of events make it very difficult to formulate any intelligent predictions about the future. Some of these, but by no means all, were mentioned in the last chapter. But there are other, perhaps even more powerful, factors which have so far only been mentioned in passing, if at all. The most significant of these, perhaps, are: population, international debt, issues of equity both within and between countries, and the greenhouse effect. Policies which are adopted in relation to these will be the ultimate determinants. It would require another book to discuss the first three of these, but some words about the last are relevant here.

There are now three ways in which a country should view its policies for its forests and its forestry. All are equally important and all are linked, although there is some conceptual advantage in considering each separately. The first is concerned with the way in which a country uses its forest land, whether for protection, production or conversion. The second has to do with the demand for wood and wood products (industrial wood for domestic use or export, and fuelwood) and how this demand is met either from natural forest or from planted trees. The third has to do with the use of forest and trees as sinks of carbon.

It seems, somewhat surprisingly, that the most successful policy for the planned use of natural forest is as "protected areas" – national parks and nature reserves. About 40 million ha of forest are said by IUCN to be under protective regimes of one kind or another, compared with the figure that we have quoted of less than 1 million ha under operational management for the sustainable production of timber. This figure, however, conceals many uncertainties. The areas have often been chosen because they are unsuitable for other uses rather than for their inherent suitability to preserve biological diversity; their management status is often dubious; and their long-term success depends, against a future of climatic uncertainty, on their internal heterogeneity (in altitude, soils, drainage characteristics, etc.) and in their being surrounded by areas of managed forest – which may not be the case. A high priority needs, therefore, to be placed on the maintenance of forest cover for a number of different purposes, only one of which is the production of timber.

Policies for wood production need to be looked at from another perspective. In the case of fuelwood, the perspective should be one of a total energy policy. When fuelwood (rather than other types of energy or agricultural residues) is deemed to be the answer, there needs to be a local source, which can be either from planted trees or, if this is possible, from sustainably managed natural forest. Industrial wood poses other problems. Its value may make long-distance transport feasible but the proper source for bulk cellulose may well be from industrial plantations, which should rather be considered as tree farming and governed by the same economic considerations as, for example, rubber or oil palm. The long-term role of the natural forest, apart from its environmental qualities and the production of those other forest products which have not been brought into cultivation, is likely (as A.J. Leslie argued at the ITTO seminars) to lie in producing high-value timbers – the Rolls Royce end of the market rather than the Ford.

It is often argued that the development of plantations is one way of saving the tropical forest; the annual increment may be more than ten times as great per hectare. But this saving role is by no means certain. It has often proved easier to establish plantations by clearing new forest than by planting previously deforested land: the soils are more fertile and there are fewer problems of land tenure. In fact there is a place for both plantations and natural-forest management; their purposes and their products are different.

The exact contribution of deforestation to the increase of carbon dioxide in the atmosphere is not yet certain, but is perhaps of the order of one-quarter of that provided by the burning of fossil fuels. The effectiveness of forest in locking up carbon depends upon the standing woody biomass, not upon the rate of growth. The highest biomass is generally contained in heavily stocked stands of old-growth trees – the stands which have been the principal target of loggers. Anything that can be done to increase the number of standing trees will make some contribution to reducing the danger of global warming, even though there is great uncertainty about the exact quantity of that contribution. A new element now needs to enter land-use and forest policies: to maximize, other things being equal, the standing volume of wood on the land. This would suggest some changes of direction for forest policies: greater emphasis on protection; less stress on harvesting stands before they become "over-mature"; strong emphasis on developing new plantations on land previously devoid of tree cover; and, wherever possible, the encouragement of tree crops in any agricultural land or waste land.

The forest policy of any country would then become the resultant of these three elements. The deforestation that is taking place at present runs counter to all of them.

It is evident that the future depends absolutely on the will and ability of countries with tropical forests to develop and apply effectively policies of sustainable and equitable land use, among them policies for the sustainable use of their forest resources. Any outside initiative should be judged by the extent to which it helps them to do so. The present cocktail of incentives and penalties, praise and strictures, which tends to flow from the better-endowed half of the world, is not a well-forged instrument for this purpose.

NOTE

A very encouraging development can be reported just as this book goes to press (July 20 1989). This is the recent initiative taken by the Government of Brazil in inviting the Overseas Development Administration of the United Kingdom to enter into collaboration with Brazilian institutions in the environmental field, and especially in the conservation and sustainable utilization of tropical forest. This has led to a memorandum of understanding between the two governments for a joint programme of research, training and scientific exchanges covering three fields of the greatest relevance to the subject of this book: the sustainable management of tropical forest for timber and other products; the conservation of genetic resources; and the monitoring of environmental influences related to climatic change.

Additional Reading

Prepared by Tropical Forestry and Computing Ltd, 93 Gidley Way, Horspath, Oxford OX9 1TQ.

Baur, George N., *The Ecological Basis of Rainforest Management* (Sydney, Australia: Forestry Commission of New South Wales, 1964).

Beusekom, C.F. van, C.P. van Goor and P. Schmidt (eds), *Wise Utilisation of Tropical Rain Forest Lands. Proceedings of the Netherlands MAB-Unesco Symposium "Tropisch regenwoud: verantwoord gebruik?" Amsterdam, Netherlands, 10-12 December 1984*, Tropenbos Scientific Series No.1 (Ede, Netherlands: Tropenbos Programme, 1987).

Bormann, F.H. and G. Berlyn (eds), "Age and growth rate of tropical trees", *Yale University Bulletin*, no. 94 (1981).

Boxman, O, N.R. de Graaf, J. Hendrison, W.B.J. Jonkers, R.L.H. Poels, P. Schmidt and R. Tjon Lim Sang, "Towards sustained timber production from tropical rain forests in Suriname", *Netherlands Journal of Agricultural Science*, 33: 125-32.

Carpenter, R.A. (ed.), *Assessing Tropical Forest Lands: Their Suitability for Sustainable Uses. Proceedings of the Conference on Forest Land Assessment and Management for Sustainable Uses, East-West Center, Hawaii, 19-28 June 1979* (Dublin, Eire: Tycooly International Publishing Co., 1981).

Caufield, Catherine, *In the Rainforest. Report from a Strange, Beautiful, Imperiled World* (New York: Alfred A. Knopf, and London: Heinemann, 1985).

Chadwick, A.C. and S.L. Sutton (eds), *Tropical Rainforest. The Leeds Symposium*. Special publication (Leeds: Leeds Philosophical and Literary Society, 1984).

Davidson, J. *et al.* "Economic use of tropical moist forests", *The Environmentalist*, 5 (supplement). (Reprinted in Commission on Ecology Papers no. 9, Gland, Switzerland: International Union for Conservation of Nature and Natural Resources, 1985).

Dawkins, H.C., *The Management of Natural Tropical High Forest with Special Reference to Uganda*, Imperial Forestry Institute Paper No. 34 (Oxford, Imperial (now Oxford) Forestry Institute, 1958). [See also the valuable consultancy reports by the same author, available in the Oxford Forestry Institute Library: Guyana (1961), Sierra Leone (1965), Ghana (1966), Sabah (1968), Sarawak (1970), Nigeria (1980)].

—— "The productivity of tropical high-forest trees and their reaction to controllable environment", D.Phil. thesis, Oxford, Commonwealth Forestry Institute, Department of Forestry, University of Oxford, 1963.

FAO, Voluntary papers submitted to the technical conference on tropical moist forests. Document FO:FDT/76/11. (Rome: Food and Agriculture Organization of the United Nations, Committee on Forest Development in the Tropics, 1976). (One hundred voluntary papers were submitted to the conference, which was cancelled. The papers cover a vast range of topics and their titles are listed in this FAO document. All are available at the Oxford Forestry Institute.)

Fox, J.E.D., "The natural vegetation of Sabah and natural regeneration of the dipterocarp forests", Ph.D. thesis, Bangor, University College of North Wales, 1972.

Fox, J.E.D. and A.J. Hepburn, *Manual of Silviculture Sabah for Use in the Productive Forest Estate*, Sabah Forest Record No. 8 (Kota Kinabalu, Sabah: Forest Department, 1972).

Furtado, J.I. (ed.), *Tropical Ecology and Development. Proceedings of the Fifth International Symposium on Tropical Ecology, Kuala Lumpur, Malaysia, 16-21 April 1979* (Kuala Lumpur, Peninsular Malaysia: International Society of Tropical Ecology and University of Malaya, 1980).

Gomez-Pompa, A, C. Vasquez Yanes, and C. Guevara, "The tropical rain forest: a non-renewable resource", *Science*, 177: 762-5.

Graaf, N.R. de, *A Silvicultural System for Natural Regeneration of Tropical Rain Forest in Suriname*, Series Ecology and Management of Tropical Rain Forests in Suriname (Wageningen, The Netherlands: Landbouwhogeschool, 1986).

Hadley, M. (ed.), "Rain forest regeneration and management. Report of a workshop at Guri, Venezuela, 24-28 November 1986", Special Issue no. 18. *Biology International* (Paris, France: International Union of Biological Scientists, 1988).

Haig, I.T., M.A. Huberman and U Aung Din (eds), *Tropical Silviculture*, FAO Forestry and Forest Products Studies No. 13 (Rome: Food and Agriculture Organization of the United Nations, 1958). "Based on a series of general invitation papers on this topic presented at the Tropical Section of the Fourth World Forestry Congress, Dehra Dun, December 1954, together with a broad survey of other pertinent literature. The appropriate invitation papers . . . are included in Volumes II and III of this study."

Haig, I.T. (ed.), *Proceedings of the Duke University Tropical Forestry Symposium* (Durham, North Carolina: School of Forestry, Duke University, 1965).

Hallsworth, E.G. (ed.), *Socio-economic Effects and Constraints on Tropical Forest Management* (Chichester: John Wiley & Sons, 1982).

Hamilton, L.S. and P.N. King (eds), *Tropical Forested Watershed – Hydrologic and Soils Response to Major Uses or Conversions* (Boulder, Colo.: Westview Press, 1983).

Jonkers, W.B.J., *Vegetation Structure, Logging Damage, and Siliviculture in a Tropical Rain Forest in Suriname*, Series Ecology and Management of Tropical Rain Forests in Suriname (Wageningen, Netherlands: Landbouwhogeschool, 1978).

Lee Hua Seng, "Silvicultural management options in the mixed dipterocarp forests of Sarawak", M.Sc. thesis, Canberra, Australia: Australian National University, 1981.

Leigh, Egbert G., Jr, A. Stanley Rand and Donald M. Windsor (eds), *The Ecology of a Tropical Forest: Seasonal Rhythms and Long-term Changes* (Washington, D.C.: Smithsonian Institution Press, 1982).

Leslie, A.J., "Where contradictory theory and practice coexist", *Unasylva*, 29(1), no. 115: 2-17, 40.

—— "A second look at the economics of natural management systems in tropical mixed forests", *Unasylva*, 39(1), no. 155: 46-58.

Leslie, A.J. (synthesizer) *Management Systems in the Tropical Moist Forests of Asia* (Rome: Food and Agriculture Organization of the United Nations, in press). (Includes contributions from C.P.S. Nair for India, Thang Hooi Chiew and Abdul Rashid bin Mat Amin for Malaysia and Bureau of Forest Development staff for the Philippines. Leslie has written the introduction and conclusions.)

McDermott, Melanie J. (ed.), *The Future of the Tropical Rain Forest*.

Proceedings of an International Conference, St Catherine's College, Oxford, 17-18 June 1988 (Oxford: Oxford Forestry Institute, 1988).

Maitre, H.F., "Natural forest management in Côte d'Ivoire", *Unasylva*, 39 (3&4), nos 157/158: 53-60.

Masson, J.L., *Management of Tropical Mixed Forests. Preliminary Assessment of Present Status*, Document FO:MISC/83/17 (Rome: Food and Agriculture Organization of the United Nations, 1983).

Mergen, François (ed.), *Tropical Forests: Utilisation and Conservation. Proceedings of an International Symposium on Tropical Forest Utilisation and Conservation, Ecological, Socio-political and Economic Problems and Potentials* (New Haven, Conn.: Yale University, School of Forestry and Environment Studies, 1981).

Mergen, François and Jeffrey R. Vincent (eds), *Natural Management of Tropical Moist Forests. Silvicultural and Management Prospects of Sustained Utilisation* (New Haven, Conn.: Yale University, School of Forestry and Environmental Studies, 1987).

Mervart, J., "Growth and mortality rates in the natural high forest of Western Nigeria", *Nigerian Forestry Information Bulletin* (Ibadan, Nigeria: Federal Department of Forest Research, 1972).

Moore, D. (synthesizer), *Intensive Multiple-use Forest Management in the Tropics. Analysis of Case Studies from India, Africa, Latin America and the Caribbean*, FAO Forestry Paper No. 55 (Rome: Food and Agriculture Organization of the United Nations, 1985).

Neil, P.E., *Problems and Opportunities in Tropical Rainforest Management*, CFI Occasional Papers No. 16 (Oxford: Commonwealth (now Oxford) Forestry Institute, 1981).

Nicholson, D.I., *The Effects of Logging and Treatment on the Mixed Dipterocarp Forests of South East Asia*, Document FO:MISC/79/8 (Rome: Food and Agriculture Organization of the United Nations, 1979).

Plumwood, V. and R. Routley, "World rain forest destruction – the social factors", *The Ecologist*, 12(1): 4-22.

Poore, M.E.D., *Ecological Guidelines for Development in Tropical Forest Areas of South-East Asia*, IUCN Occasional Paper No. 10 and IUCN Publications New Series No. 32 (Morges, Switzerland: International Union for Conservation of Nature, 1975).

—— *Ecological Guidelines for Development in Tropical Rain Forests*

(Morges, Switzerland: International Union for Conservation of Nature, 1976).

Poore, Duncan and Jeffrey Sayer, *The Management of Tropical Moist Forest Lands: Ecological Guidelines. IUCN Tropical Forest Programme* (Gland, Switzerland: International Union for the Conservation of Nature and Natural Resources, 1987).

Richards, P.W., *The Tropical Rain Forest. An Ecological Study* (Cambridge: Cambridge University Press, 1952).

Routley, R. and V. Routley, *The Fight for the Forest* (Canberra, Australia: Australian National University Press, 1975).

Saulei, S. (ed.), *Tropical Rain Forest Ecology and Management. Report on a Sub-regional Workshop at Lae, Papua New Guinea, 20 June-1 July 1988* (Lae, Papua New Guinea: University of Papua New Guinea, Biology Department, 1988).

Schmidt, R.B., "Tropical rain forest management – a status report", *Unasylva*, 39(1), no. 156: 2-17.

Schmithusen, F., *Forest Utilisation Contracts on Public Land*, FAO Forestry Papers No. 1 (Rome: Food and Agriculture Organization of the United Nations, 1977).

Shepherd, K.R. and H.V. Richter (eds), *Managing the Tropical Forest. Papers from a Workshop Held at Gympie, Australia, from 11 July to 12 August 1983* (Canberra, Australia: Development Studies Group, The Australian National University, 1985).

Shugart, H.H., M.S. Hopkins, I.P. Burgess and A.J. Mortlock, "The development of a succession model for sub-tropical rainforest and its application to assess the effects of timber harvest at Wiangaree State Forest, New South Wales", *Journal of Environmental Management*, 11: 243-65.

Spears, J.S., "Can the wet tropical forest survive?", *Commonwealth Forestry Review*, 58(3): 165-80.

Suparto, Rahardjo S. and seven others (eds), *Proceedings of a Symposium on the Long-term Effects of Logging in Southeast Asia, Damarga, Bogor, Indonesia, 24-27 June 1975*, BIOTROP Special Publication No. 3 (Bogor, Indonesia: SEAMEO Regional Centre for Tropical Biology (BIOTROP), 1975).

Sutton, S.L., T.C. Whitmore and A.C. Chadwick (eds), *Tropical Rain Forest: Ecology and Management*. Special publications series of the British Ecological Society No. 2 (Oxford: Blackwell Scientific Publications, 1983). (Papers presented at the Leeds Symposium in 1982.)

Synnott, T.J., *Tropical Rain Forest Silviculture: a Research Project*

Report, CFI Occasional Paper No. 10 (Oxford: Commonwealth (now Oxford) Forestry Institute, 1979).

Synnott, T.J. and R.H. Kemp, "Choosing the best silvicultural system", *Unasylva*, 28(1), nos 112-13: 74-9. (Condensed from "The relative merits of natural regeneration, enrichment planting and conversion techniques", Document FO:FDT/76/7a (Rome: Food and Agriculture Organization of the United Nations, Committee on Forest Development in the Tropics, fourth session, 15-20 November 1976).)

UNESCO/UNEP/FAO, *Tropical Forest Ecosystems. A State-of-knowledge Report*, ed. A. Sasson and B. Hopkins, Natural Resources Research No. 14 (Paris: United Nations Educational, Scientific and Cultural Organization: 1978).

US Interagency Task Force on Tropical Forests, *The World's Tropical Forests: a Policy, Strategy and Program for the United States*, Department of State Publication No. 9117 (Washington, DC: US Government Printing Office, 1980).

Wadsworth, F.H., "Forest management in the Luquillo Mountains. I. The setting", *Caribbean Forester*, 12(3) (1951),: 93-114.

—— "Forest management in the Luquillo Mountains. II. Planning for multiple land use", *Caribbean Forester*, 13(2) (1952a),: 49-61.

—— "Forest management in the Luquillo Mountains. III. Selection of productions and silvicultural policies", *Caribbean Forester*, 13(3) (1952b),: 93-119.

—— (in press, a major review and synthesis of silviculture and management in tropical rain forests. This work was to be issued to coincide with the golden jubilee celebration of the Institute of Tropical Forestry, San Juan, Puerto Rico in May 1989).

Wan Razali bin Wan Mohammad (ed.), *Proceedings of the IUFRO S4.01 and S4.02 Joint Meeting on Growth and Yield in Tropical Mixed/Moist Forests. Kuala Lumpur, Malaysia, 20-24 June 1988* (Kepong, Peninsular Malaysia: Forest Research Institute, Malaysia, 1989).

Webb, L.J., *Ecological Considerations and Safeguards in the Modern Use of Tropical Lowland Rainforests as a Source of Pulpwood: Example of the Madang Area in Papua New Guinea* (Waigani, Papua New Guinea: Office of Environment and Conservation, 1977).

Willan, R.G. (synthesizer), "Management systems in the tropical moist forests of Africa" (Probably will be published in the series FAO Forestry Papers) (Rome: Food and Agriculture Organization of the United Nations, in press). (This document will be a synthesis of two compilations, the first by R. Catinot for francophone Africa and the second by M.S. Philip for anglophone Africa. The draft version (1988) of the anglophone section includes contributions by P.R.O. Kio from Nigeria and P.K. Karani from Uganda.)

Wyatt-Smith, J., *Manual of Malayan Silviculture for Inland Forests*, Malayan Forest Records No. 23 (Kuala Lumpur, Peninsular Malaysia: Forest Department, 1963).

—— *The Management of Tropical Moist Forest for the Sustained Production of Timber: Some Issues*, IUCN/IIED Tropical Forest Policy Paper No. 4 (London: International Institute for Environment and Development, Forestry and Land Use Programme, 1987).

Index

Abidjan Resolution (1989), 224–5

Africa: tropical moist forest (TMF) in six ITTO countries (Cameroon, Congo, Côte d'Ivoire, Gabon, Ghana, Liberia)
cutting cycles, 54–6
deforestation, 44, 48
economic circumstances, 64–6, 71–2
environmental considerations, 46, 49
fees and taxes, 60–62, 72
forestry services, 62–3, 71
in–country processing, 59–61, 72
inventories, 66–7
lack of domestic markets, 47–8, 58–9, 204
LKS promotion, 47, 58–9, 62, 72
logging, 48, 53–8, 62–4, 71–2
management for sustained yield, 43–4, 53–4, 70–73, 194
non–industrial goods/ benefits, 40–41
plantations, 41–2, 70–71
policy/laws, 43, 48–53, 58–9
research, 63, 65–7, 72
reservation, 50–52
silviculture, 41, 43–4, 67–70
training programmes, 64

agricultural pressure on forest lands in ITTO member countries
Africa, 44, 46–7, 49–51, 71
Asia, 117–18, 140, 145, 151, 197
South America/Caribbean, 77–80, 82–4, 89, 102, 105, 108, 198
see also deforestation; demography; farmers

agro-ecosystem, 16

agroforestry, 10, 104–6

Amélioration des Peuplements Naturels (APN), 41, 43, 68–9

annual allowable cut (AAC), 5–6, 207
in African TMF, 45–6
in Queensland rain forests, 33–4
see also cutting cycles

Arena Tropical Shelterwood System, 94–5

Asia, forest areas in five ITTO countries (Indonesia, Malaysia, Papua–New Guinea, Philippines, Thailand)
concession agreements, 47, 126–32